Foreword

Considerable changes have taken place in farming this century, particularly since the end of the Second World War, but because most of the population live in towns and cities, some of these changed have gone largely unnoticed. Although many people are interested in farming they are often not very well informed about it, and have a sentimental view of rural life which bears little relationship to the practical realities of farming in the 1980s.

This book has been written for young people who want to find out more about farming and sets out to give a realistic and up to date account of agriculture in the United Kingdom. To do this, farming routines are described whenever possible, but in addition the reasons behind them are explained, as are the ways that traditional practices have been altered by advances in modern technology. I hope that in this way the reader will not only gain an idea of the practicalities of modern farming, but also some understanding of its theory and its problems.

CHAPTER ONE

Farming in the United Kingdom

This century has seen a dramatic change in the face of farming in the United Kingdom as a result of scientific and technological advances. Sixty years ago most jobs on the farm were done by hand or with horses, crop yields were much lower and disease caused many problems in farm animals. Over that sixty years a whole range of farm machinery has been developed which has replaced not only the horse, but many farm labourers as well. Advances in plant and animal breeding have given the farmer new, more productive strains and crop yields have increased considerably because of the use of artificial fertilisers and chemicals to kill weeds and crop pests. At the same time, developments in veterinary medicine have helped to control disease in animals, and the use of modern building materials, together with improvements in the design of farm buildings have allowed larger numbers of animals to be kept housed.

Taken together these changes have produced considerable increases in yield and farm output. Figure 1.1 shows the increase in agricultural output over a ten-year period (1969-1979) for four types of produce — farm crops, horticultural produce, livestock and livestock products, such as milk and wool. The output is a measure of the cash value of the products, which has been adjusted to remove the effects of inflation on the prices.

The greatest increase in production occurred in the middle of this century, as can be seen by the fact that between 1937 and 1957 the total farming output increased by a factor of almost five. This increase in production has

2 Farming Today

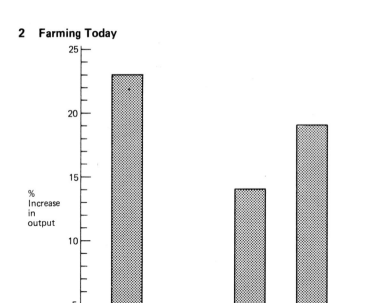

Figure 1.1 Percentage increases in agricultural output over the ten years 1969-1979.

taken place at a time when the number of people working on the land has steadily declined, and there has been a continual loss of farming land as towns and cities have become bigger. Over the past few years there has been an average loss of about 15,000 hectares (60,000 acres) of crops and grass each year and in the fifty-year period, 1927-1977, this area has fallen by 8.4% — a loss of almost one tenth of the most productive agricultural land.

The increase in food production, so important to feed an increasing population, has not been achieved without cost. Many people are concerned about the effects of agricultural chemicals on the environment, and about the welfare of

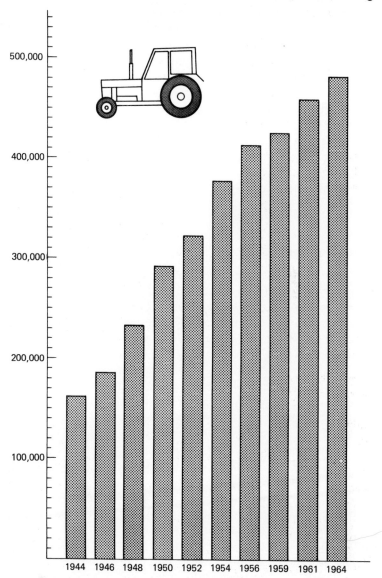

Figure 1.2 Numbers of agricultural tractors in Great Britain (1944-1964).

4 Farming Today

farm animals kept under intensive conditions, and these problems will be considered later in the book. At the same time the appearance of the countryside has changed, for the worse in many people's opinion. One obvious difference in many areas is the loss of hedgerows, which have been pulled up not only to make it easier to use machinery, but also to cut out the cost of hedge trimming and maintenance. However, it should be remembered that there is no such thing as a 'typical' country scene, because the rural landscape is entirely man-made, the result of 2,000 years of agricultural change. If it were not for farming activities in the past, most of Great Britain would be covered with woodland.

Mechanisation

As we have seen there have been many changes in farming, but without doubt the most important has been the development and use of machines. The steam engines of the nineteenth century were replaced by the tractor, powered by an internal combustion engine early in this century, and since then tractors have become larger and more powerful with an increasing range of implements that can be attached to them.

Machines can quickly do jobs that in the past took many man-hours and therefore they have tended to replace farm workers. Figure 1.2 shows the numbers of tractors on British farms over a twenty-year period (1944-1964). Since the mid-sixties the number of tractors has remained steady or even gone down slightly although they have tended to become more powerful.

Figure 1.3 shows the fall in people working on the farm between 1948 and 1968, and a comparison with Figure 1.2 shows that tractors numbers increased as farm workers decreased. The use of machines on the farm has therefore tended to speed up the drift of the population away from the country to the towns and cities, which has been going on since the Industrial Revolution.

Another effect of mechanisation has been to increase

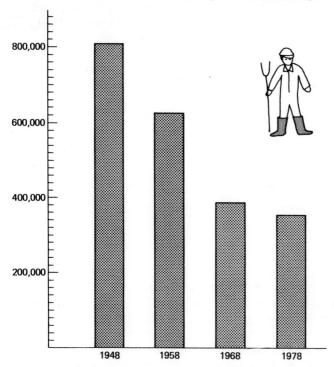

Figure 1.3 Numbers of people employed in agriculture and horticulture (1948-78).

farm size, as small farms have been amalgamated into larger units, because by using machines one man can farm a far greater area than previously. Figure 1.4 compares the numbers of small (less than 50 hectares) and large farms in 1967 and 1977.

Types of farms

Compared with fifty years ago, a typical farm in the United Kingdom is larger, highly mechanised and has a much smaller workforce. However, the advances in technology

6 Farming Today

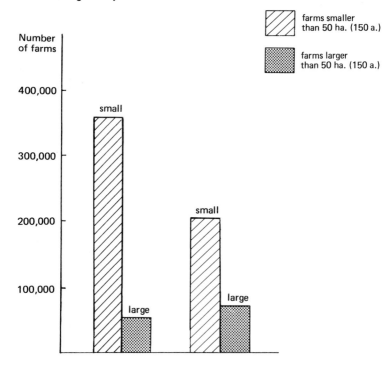

Figure 1.4 A comparison of farm size in 1967 and 1977.

that we have looked at have in no way produced a 'standard' farm. Many important things which affect the farmer, such as the weather, the height of the farm above sea level and the type of soil cannot be changed and these will influence the type of farm and in particular the crops that are grown. Even within one part of the country each farm will differ from its neighbour, though it is usually possible to make generalisations about 'typical' farms in an area.

In general terms we can say that:

1) Dairy farms are found mainly in the west of the country where the high rainfall gives good grass growth for the cows to produce milk.

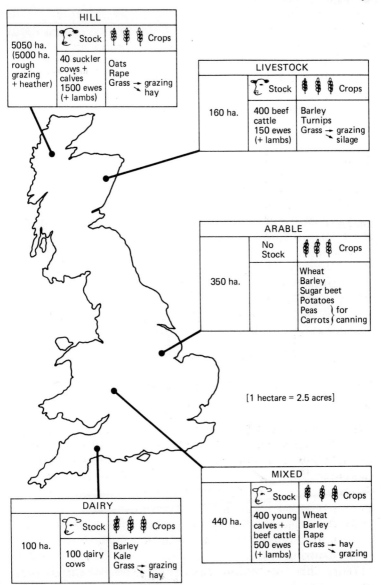

Figure 1.5 Some typical farms

2) Arable farms (farms with few or no livestock) are found mainly in the flat eastern part of the country, particularly in the south where the summers are warm and there is relatively low rainfall.

3) Livestock farms are found in the cooler north and western areas where the conditions are less favourable. Some hill farms have such poor land that only hardy sheep are kept there.

Despite these broad groupings, most farms are mixed, growing several different crops for fodder or for sale and keeping some livestock. Figure 1.5 shows the main crops and stock on five typical farms in different parts of the country. These are:

1) A Scottish hill farm which is by far the largest but also the least productive because over 98% of the land is rough moorland.

2) A livestock farm in north east Scotland where the crops are grown mainly to feed the beef cattle and the sheep.

3) An arable farm on the rich East Anglia soil with no livestock and crops grown for sale off the farm, including vegetables for nearby canning factories.

4) A dairy farm in the West country, an area of high rainfall and good grass growth that also grows barley and kale as winter feed for the cows.

5) A mixed farm in the Midlands with young calves, beef cattle and sheep that grows a range of crops for sale or for feeding.

Two other things will influence the type of farm, firstly the interests and experience of the farmer and secondly economic factors. A farmer is running a business, so it is important to make a profit and this will influence the crops grown or stock kept. For example, in recent years cereal prices have risen, and to take advantage of this, greater areas of corn have been sown, some in wetter upland areas where cereal growing would not have been profitable in the past.

From this we can see that there is no such thing as a typical farm but certain types of farming are more likely to be found in certain parts of the country. Although climate and soil type are very important in determining what is grown where, the choice will also be influenced by economic factors.

CHAPTER TWO
Dairy Farming

For most people, milk and other dairy products are an important part of the diet. On average, each person drinks 2½ litres (4½ pints) of milk and eats 150 g (5¼ oz) of butter and cheese every week, and dairy products provide the average household with much of their protein. For the farmer, milk is one of the major products of agriculture, although the number of dairy farms varies from one part of the country to another. Figure 2.1 is a map of England and Wales showing the number of dairy cows found in different regions. The best areas are lowlands, with a high rainfall, giving lush grass growth to feed the cows, such as are found in the Cheshire Plain and West Country. Northern Ireland, which has high rainfall is quite important for dairying, but Scotland is much less so.

The dairy cow

Like all mammals, the dairy cow produces milk as a food for her young, but unlike most of them, milk is produced in amounts far greater than the needs of the calf. This is due to selective breeding, because farmers have bred from cows with a high milk yield, so that over a long period of time the modern dairy cow has been produced. Cows yield on average 4,600 litres (1,000 gallons) per year which is equivalent to filling 28 pint milk bottles a day.

Over the years there has been a dramatic increase in milk yields from British herds, brought about by improved breeding and management of the cows. (Figure 2.2.)

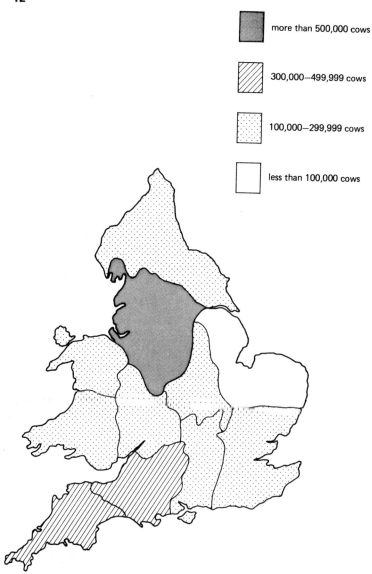

Figure 2.1 The total number of dairy cows in each Milk Marketing Board region of England and Wales.

Figure 2.2 Average annual milk yields in England and Wales.

The beef cow is also a product of selective breeding, but in this case meat production is the desirable quality. Compare the two cows shown in the photographs overleaf (Figures 2.3 and 2.6). The dairy cow looks quite thin and bony, not because she has been badly fed, but because the food she eats is used to make milk. In contrast, the beef cow looks much sleeker, with a good flesh cover, but a smaller udder.

Breeds

More than 80% of the dairy cows in England and Wales are the black and white Friesians, which originally came from the Netherlands. They have a high milk yield, and their calves are also suitable for fattening to produce beef.

Red and white Ayrshire cattle are popular in Scotland, but although their milk yield is second only to the Friesian, their calves are less suitable for beef. They may have horns but these are removed when they are very young because of the damage that they can do to other cows.

The fawn-coloured Channel Island breeds (Guernsey and Jersey) originated, as their name suggests in the Channel Islands. Although they produce a rich creamy milk, they are much less popular nowadays because their yield is lower than that of the Friesian or Ayrshire, and their calves are not suitable for meat production.

Figure 2.3 A beef cow (Aberdeen-Angus Cattle Society).

Figure 2.4 A Friesian cow.

Figure 2.5 An Ayrshire cow (**Ayrshire Cattle Society**).

Figure 2.6 A Jersey cow (**Jersey Cattle Society**).

The Dairy Shorthorn was once the major milk producer in Britain, but it has now almost disappeared, except in Northern Ireland. The coat colour is variable, being a mixture of red, red roan, and white.

The composition of milk

An average milk sample is made up of 87.5% water and 12.5% solids (Figure 2.7). The milk solids are fat, protein, milk sugar, vitamins and minerals. At the turn of the century, minimum levels for milk solids were laid down by law to stop people from watering down the milk, and even today the farmer is paid more if the milk has a higher percentage of solids.

Breeding

Every 20 days a cow that is not pregnant has a period of *oestrus* or heat, which lasts for 18 hours and this is the only time that she will mate. It is followed 12 hours later by ovulation (or release of eggs). The farmer will watch the herd for cows that are "bulling", that is mounting or being mounted by other cows, so that service can be arranged when they are most likely to conceive.

Cows can be served naturally by the bull or by *artificial insemination* (A.I.). For A.I., semen is collected from good quality bulls, which are kept at special A.I. Centres. This is normally done by putting a cow called a "teaser" in a specially designed service pen. When the bull mounts the cow, his penis is directed into a tube in which the semen is collected. The semen is then examined under the microscope to check that the sperm appear healthy, then diluted and frozen in liquid nitrogen at $-193°C$, so that it can be kept for several years. When a cow comes on heat, a sample of semen is unfrozen, taken to the farm and injected into the cow's vagina using a long tube. A. I. has the advantage to the farmer that the cows can have calves by a high quality bull without the expense or the danger involved in

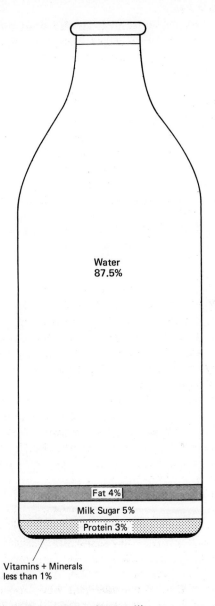

Figure 2.7 Average composition of raw milk.

keeping the bull. This is particularly important with dairy bulls which can be much more aggressive than bulls of the beef breeds.

If a bull is kept, it is usually housed, and mating is controlled, that is the cows are only taken to the bull when they are on heat, although sometimes a bull will be left to run in a field with a group of heifers (young females which have not had a calf). The breed of bull used will depend on the type of calf to be bred. For heifers it is important that the bull should sire a small bodied calf, to make calving easier. If the calves are to be fattened for beef, a beef bull such as the Hereford may be used to produce cross-bred calves, but if the aim is to breed replacements for the dairy herd, a dairy bull of the same breed is needed. In this case, of course only the female calves are kept and the males will be fattened. One of the reasons for the great increase in popularity of the Friesian since the war is that even purebred animals produce good beef carcasses.

The average length of pregnancy is 282 days or about nine months, and normally only one calf is born, with the outstretched front legs appearing first and the head resting on them. Sometimes the calf is not lying this way, in which case the cow will need help in calving from either the farmer or the vet. The calf is removed from the mother shortly after birth and reared artificially so that the cow's milk can be sold, but it is very important for the calf to be given *colostrum,* which is the first milk produced by the cow after calving. This is much richer in protein than normal milk and contains antibodies which help to protect the new born calf from many diseases. Calves quickly learn to drink from a bucket and are fed on milk or milk substitute for four to six weeks, and then gradually weaned onto solid food.

Lactation or milk production

After calving, milk production increases to a peak at three to six weeks and then slowly declines. The actual yield will vary from one cow to another depending on such factors as

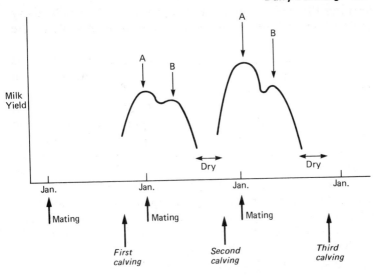

Figure 2.8 Milk yield of one cow over a three-year period.

breed, feeding, age of the cow and time of year.

As we have already seen, a cow produces milk because she has had a calf, and it therefore follows that to go on making milk for the farmer, she must have a calf at regular intervals. She will come back on heat between five and ten weeks after calving and should be mated shortly after this. Most farmers aim for a calving cycle of a year, so that every cow has a calf every 12 months, and this means that for the greater part of each lactation the cow is carrying another calf. Not suprisingly, milk yield falls off towards the end of pregnancy, and it is normal practice to 'dry off' the cow, so that there is a period of six to nine weeks, without milking before the next calf is born. This is done by reducing the feed and the frequency of milking until milk production stops. Great care is needed as the cow is particularly vulnerable to *mastitis* (infection of the udder) at this time. In the last weeks of pregnancy, the cow is 'steamed up' or generously fed to make sure she is in peak condition for calving and the subsequent milking.

Figure 2.8 shows a pattern of lactation for a heifer mated in January at about 18 months or older so that she calves in October. The following January she is mated again, and is dried off about eight weeks before the second calf is born. Notice that these are two peaks of production, one (marked A on the graph) shortly after the calf is born, and the second lower one (B) in the spring, coinciding with the grass growth. The graph also shows that more milk is produced in the second lactation, and in fact yield increases with each calving up to the fourth.

Farmers specialise in either winter milk production with cows calving in the autumn, or summer milk production with a spring calving herd.

The year on the dairy farm

On the farm there will be a yearly cycle associated with the time of calving, and the activities involved in growing fodder for the cows. Figure 2.9 summarises the changes that take place month by month on a farm with an autumn calving herd that is fed on kale, silage and barley in winter. The times given are approximate and will vary from one part of the country to another, and in practice, calving is spread out for several months after October because the cows do not always conceive on first mating.

The outer circle shows the reproductive and lactation cycles of the herd while the inner one relates to their feeding.

Feeding for milk production

The cow can only make milk if she has an ample supply of the necessary nutrients or foodstuffs and these will be supplied either from her own body reserves or from the food that she eats. A high yielding cow does tend to lose weight in the early stages of lactation and this means that

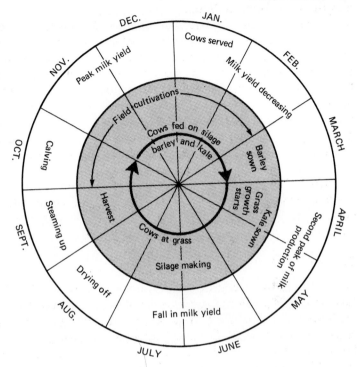

Figure 2.9 The year on a dairy farm.

some of her body reserves are being broken down to aid milk production. However, these reserves are limited and therefore, the cow must be fed an adequate diet to keep up a good yield. In general terms the more milk a cow produces, the more food she will need to replace the nutrients that have been used to make the milk, so that cows with a high milk yield will need more food than those with a low milk yield. In addition, a cow will be given different amounts depending on its stage in the milk production cycle. For example, in the weeks after calving milk production is high and extra nutrients are needed, whereas less food is needed as milk production falls before the birth of the next calf.

The composition of the milk can be affected by the diet

22 Farming Today

of the cow, so that cows in poor condition may produce milk with a lower than average amount of milk solids. This is important to the farmer because payment for the milk is based on its composition.

Figure 2.10 A modern parlour (Gascoigne Milking Equipment Ltd).

There are different systems of feeding dairy cows and there are ways of calculating the rations needed to make sure that the food provides sufficient energy, together with protein and other nutrients for milk production. In most cases feeding is based on: *bulky foods*, grass in summer, hay or silage in winter and *concentrates* which are high quality foods based on cereals with added protein and minerals.

Feeding costs are high, so the calculation of the rations is an important part of the dairy farmer's job. If each cow is treated as an individual and the diet tailored to her needs at any one time, the farmer should be able to obtain maximum milk production without giving more food than is needed, and this will of course increase his profits.

The dairy parlour

Modern dairy parlours are divided into stalls arranged around a sunken pit where the milker works. The cows wait outside in a collecting yard, then enter the stalls in turn and leave after they have been milked. One such parlour is shown in Figure 2.10. The cows stand with their heads to the wall, so that the milking machines (which are at the sides of the pit) can be easily attached by the person working in the pit.

Compare this parlour with the traditional tied byre shown in Figure 2.11. Cows would have been milked like this in the days of hand milking, and even with a portable milking machine it is obvious how much slower and less convenient this system is.

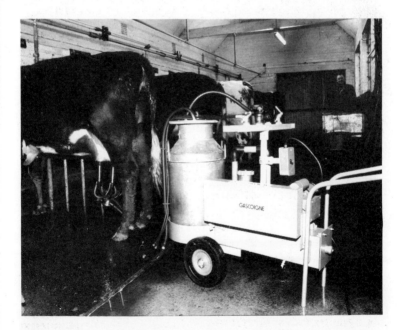

Figure 2.11 A traditional milking shed (**Gascoigne Milking Equipment Ltd**).

24 Farming Today

On a few farms there are rotary parlours where the cows enter the stalls on a circular platform which rotates around a central pit where milking takes place.

Milking

Once the cows are in the stalls, the milker removes a small amount of milk, called the fore milk, by hand and examines it for clots or blood which would indicate mastitis. Then the udder is washed to clean it and also to stimulate the 'letdown' or release of milk. Next, the milking machine consisting of four teat cups, is attached, which draws milk out of the udder by regular squeezing of the teats. Milk from each machine is usually collected in a graduated glass jar so that the yield of each cow can be recorded. Figure 2.12 shows a view of the cows from the pit, with the milking machines

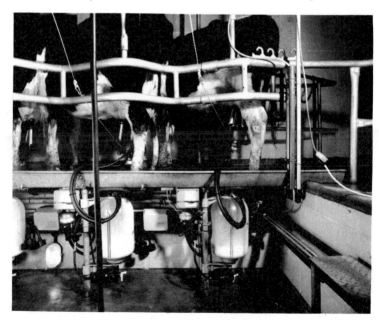

Figure 2.12 Milking **(Gascoigne Milking Equipment Ltd)**.

in position and the milk collecting in the jars.

During milking a ration of concentrates may be automatically dispensed to each cow with the amount given being related to the yield of the cow. At the end of milking the teat cups are removed by hand (or automatically in the most modern parlours), the teats are dipped in disinfectant and the cows can then leave the parlour. The time taken to milk a herd will vary but one person should be able to milk between 60 and 120 cows per hour, depending on the parlour design.

The milk flows from the jars along a pipe, through a filter and into the bulk storage tank, where it must be rapidly cooled, and stored at about 5°C until collection. Since 1979, collection of milk in churns has been discontinued and all milk is now collected daily by tanker (Fig. 2.13).

Times of milking

After each milking, more milk is secreted into the udder,

Figure 2.13 A milk tanker making a farm collection (**Milk Marketing Board**).

causing a build up of pressure which will in time slow down or even stop the production of more milk. This makes it essential to remove the milk at regular intervals. In theory, the cows should be milked at exactly 12 hour intervals but because this would mean very unsocial working hours, there is usually a ten-hour interval in the day and a 14-hour interval through the night.

The day on a dairy farm

The routine on a dairy farm is dominated by the times of milking and as Figure 2.14 demonstrates the hours worked do not correspond with most people's idea of a normal working day. The diagram shows a 24-hour period on a farm with 200 Friesian cows milked in a 16 stall herringbone parlour. Here milking starts at 3.45 in the morning and 2 o'clock in the afternoon, although in many parts of the country 5 a.m. and 3 p.m. are more common times.

Other work, apart from milking will vary with the time of year. In summer the herd will be at grass and must be collected from and returned to the field twice a day at milking time. In winter, the cows are housed and will need feeding, while in spring and autumn they will be turned out during the day and brought in at night. They must also be checked for signs of bulling and either taken to the bull or served by A.I. In addition, there are the replacement heifers which will join the herd once they have had their first calf, and the 'dry' cows. These are cows that are not being milked because they are close to calving and which must be carefully watched so that they can be helped if they have any problems whilst calving.

Hygiene

Because milk is an ideal place for bacteria to grow and multiply, conditions of scrupulous hygiene must be observed at all times. This means that the parlour must be kept

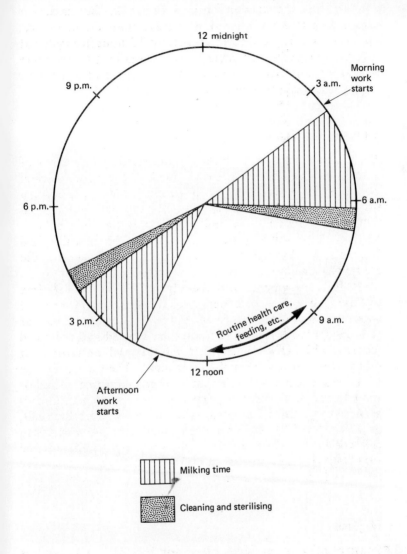

Figure 2.14 Twenty-four hours on a dairy farm

28 Farming Today

clean, and that the cows must be healthy. Particular attention is paid to the udder, which can easily become infected, and is therefore washed before milking and the teats dipped in disinfectant afterwards. In addition, all the milking equipment must be sterilised after each milking to stop the build up of deposits of milk where germs can multiply.

The way in which milk is cooled and stored is governed by Milk Marketing Board regulations aimed at keeping the milk clean and free from bacteria.

At the Dairy

Milk is tested for composition and cleanliness, and the rate of payment to the farmer is affected by these results. More than half of the milk produced is bottled (Figure 2.16) and sold as liquid milk, while the remainder is used for the manufacture of milk products, chiefly butter and cheese.

Figure 2.15 Pasteurisation or raw milk to kill harmful bacteria **(Milk Marketing Board)**.

Figure 2.16 An automatic bottling plant (Milk Marketing Board).

Before bottling the milk is pasteurised, or heat treated to kill any harmful bacteria which may be present (Figure 2.16).

The Milk Marketing Boards

During the 1930's, a series of Milk Marketing Boards (MMBs) were set up, which buy and control the sale of all the milk produced in the U.K. In addition, they control the fleet of bulk milk tankers and are therefore responsible for the collection and transport of 16.2 million litres (3.6 million gallons) of milk every day from the farm to the dairy.

All dairy farmers must be registered with the Boards, and the great majority sell their milk directly to them, although a few have their own bottling plants and sell direct to the public under licences from the Board.

CHAPTER THREE
Beef Production

An ideal beef animal is one that grows rapidly to its full size, with a good carcass that is free from too much fat, and with well developed hindquarters and back because, as shown in Figure 3.1, it is from here that the steaks and prime roasting joints are cut.

There are many breeds that are used only for beef production, but as we have seen in Chapter 2 many beef animals are either male calves from dairy cows such as the Friesian, or are produced by mating a dairy cow to a beef

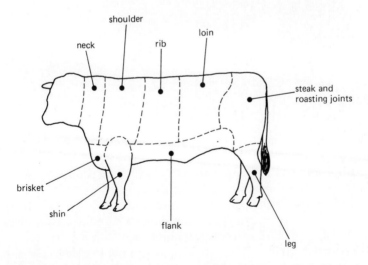

Figure 3.1 A beef carcase.

bull. In recent years 'exotic' breeds imported from the Continent have tended to replace some native breeds.

Beef breeds

Aberdeen Angus are small black cattle famous for producing high quality beef, although they are less popular today because of their size and slow growth rate. Angus bulls are

Figure 3.2 An Aberdeen Angus bull (**Aberdeen Angus Cattle Society**).

still often used to cross with dairy heifers to produce a small calf and therefore an easy first calving.

The most important of the native breeds is the Hereford, which has a good growth rate and carcase quality. Herefords are red, with a characteristic white face that is inherited by cross-bred calves. Calves produced from a Friesian cow and a Hereford bull are black with a white face and other

Beef Production 33

Figure 3.3 A Hereford bull **(Hereford Herd Book Society)**.

markings, and are one of the most popular crosses with beef producers because of their rapid growth rate.

The Charolais is a large cream or white animal which has been introduced into the U.K from France and has become very popular in recent years. The calves are large and grow rapidly, but their size at birth can lead to more calving difficulties. Charolais cross calves are easily identified by their unusual cream or creamy brown colour.

The Simmental originated in Switzerland. Because Simmentals have a large body size together with a high milk yield, they are classified as dual purpose cattle for both beef and dairying and may be used to produce cross-bred cows for suckler beef herds. Figure 3.5 shows the variation of colour, which can be from yellow to red-brown, and of marking which can be found with this breed.

There are many other native breeds, such as the Devon, Welsh Black and Shorthorn, but they are being increasingly replaced by breeds with large body size and more rapid growth rate, although as we shall see they are still used to produce cows for suckler herds.

34 Farming today

Figure 3.4 A Charolais bull.

Figure 3.5 A group of young Simmental bulls (**British Simmental Cattle Society**).

Systems of calf rearing

Most beef calves are produced by dairy cows, and must be reared artificially. As explained in Chapter 2, after receiving the colostrum or first milk, they are bucket fed with a milk substitute and are weaned onto a solid diet between four and eight weeks.

Young calves are prone to a variety of infections and as many as six out of every hundred calves may die before the age of six months, while many others grow slowly because of less severe infections. The main health problem is caused by a bacterium called *Eschericia coli* which causes scouring (diarrhoea), fever and frequently death. Calves which have not been given colostrum have no immunity to the infection, and they are particularly likely to become ill if they have been overfed with milk or kept in unsuitable conditions. Cold, damp and draughty conditions may also make the calves liable to pneumonia, which can again be fatal.

A smaller number of calves are naturally reared in *suckler herds* where they are left to suckle their own mothers, until they are weaned at about six to nine months. There are also systems of multiple suckling where a cow with a high milk yield suckles as many as four calves but the main problem here is to persuade the cow to accept the foster calves.

Most suckler herds are kept outside throughout the year, and are often kept on marginal land which is unsuitable for intensive farming. The cows must therefore be hardy and may be crosses of the traditional breeds such as the Shorthorn, which combine hardiness with suitable carcase quality.

Veal production

Compared with the Continent comparatively few veal calves are reared in the U.K. Veal is produced by rearing calves on a diet of liquid milk until they are slaughtered at 15 weeks old, and this method of feeding produces the characteristic pale meat colour. Unfortunately, it also makes the calves very liable to disease, so that they are traditionally kept in individual pens with slatted floors to reduce the risk of infection. These unnatural conditions have led to considerable criticism from the animal welfare lobby and many people refuse to eat veal because of the methods used to produce it. However, recently a British firm has pioneered a new method of rearing veal which it is claimed, is not only more humane, but also more profitable for the farmer. Under this system groups of about twenty calves are kept in strawed pens and fed from an automatic milk dispenser. These dispensers, which have teats from which the calves can suck at any time, can be seen between the two end pens on either side of the gangway in Figure 3.6. This method, which seems to form an acceptable solution to the problems of veal production, has been accepted as such by the Royal Society for the Prevention of Cruelty to Animals.

Figure 3.6 Loose housed veal calves **(Volac Ltd)**.

Castration of bull calves

A bull calf grows more rapidly than a castrated animal (usually called a bullock or a steer) but most male calves are castrated during the first few months of life because of the

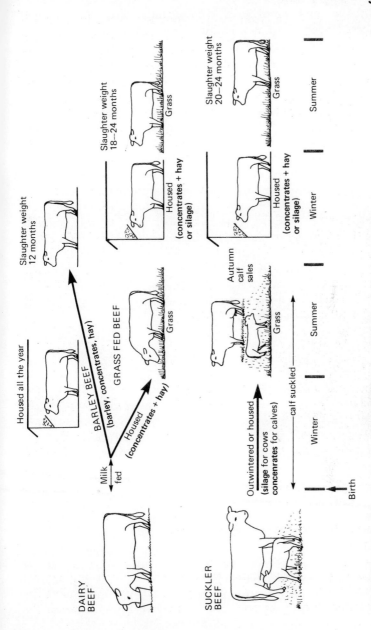

Figure 3.7 Three systems of beef production for autumn calving herds.

problems associated with keeping bulls. Not only are they more dangerous to handle, but there is always the possibility of unwanted pregnancies if they escape and get in with female animals. In addition many butchers do not want to buy bull beef which may taste stronger, and can develop an unpleasant 'taint' in older animals, although some producers now specialise in the intensive fattening of bull calves for slaughter at an early age before the taint develops.

Castration is usually carried out before the calf is three months old. There are three main methods: by surgery, by the use of rubber rings and by the Burdizzo method. The surgical method involves two small cuts being made in the scrotum, so that the testes can be pulled out and removed. The second method, using a rubber ring, is only suitable for very young calves. A tight rubber ring is applied above the testes blocking their blood supply, so that the tissue below the ring shrivels and falls off after two to three weeks. The Burdizzo castrator is a pincer-like instrument which is used to crush the spermatic cords above the testes, which will shrivel within a few weeks.

Systems of beef production

It is very difficult to generalise about how long it will take to fatten a calf to slaughter weight. This will depend on the breed and individual animal and the quality and quantity of feeding as well as the system of rearing that is used. Three common systems are: barley, beef, grass fed beef and suckler beef. The main points of each are outlined in Figure 3.7.

Barley beef

Here calves from dairy cows are bucket fed for the first weeks of life and then fed on a high quality diet consisting mainly of barley, so that they reach slaughter weight at about 12 months old. They are never turned out to grass, so food costs are high and to be successful animals with a rapid growth rate must be used.

Grass-fed beef

This is probably the most common system of fattening dairy calves, which are artificially reared on milk substitute and then fed on concentrates and hay. Calves born in the autumn will be turned out to grass in the early summer, and then brought in the following winter. At the end of the winter they will be turned out to grass again and should reach slaughter weight at some time during this second summer.

Suckler beef

As explained earlier in the chapter these calves are produced from suckler cows, not from the dairy herd and allowed to suckle thier mothers naturally. The cows usually calve outside to reduce the risk of disease and if they are born in the autumn they may be outwintered, or in exposed areas brought inside for the winter. The cows and calves are turned out in the spring, then weaned at some time during the summer, before the calves are sold at the autumn sales. Their treatment after this is usually similar to that of the grass fed beef in that they are housed in the winter and fattened off the grass towards the end of their second summer.

CHAPTER FOUR

Sheep

Historically, the major role of the sheep was as a producer of wool and from the twelfth to the nineteenth century, wool was the most important British industry. Since the early eighteenth century there has been an increasing need for meat production, and to satisfy the needs of the increasing population, sheep today are farmed for meat and their wool is only a secondary product.

Their importance in modern British agriculture is mainly because they can survive in upland areas which are unsuitable for other types of farming. They can also be kept on lowland farms but they are less profitable than other enterprises, so that there tend to be fewer flocks. Figure 4.1 is a map of the British Isles showing the distribution of sheep in Britain. From this you can see that by far the greatest number are kept in the mountains and upland areas, with the exception of Kent, the home of the Romney Marsh sheep which graze the coastal marshes.

As we have seen, sheep in Britain are kept for meat, with wool as a by-product but in some European countries, notably France and Italy, sheep are kept for their milk. In Italy, for example, milk products, mainly in the form of cheese, account for more than half of the total sales from sheep.

Breeds

There are about 40 breeds of British sheep (classified by the appearance of their wool and by the place where they live).

Figure 4.1 Sheep in Britain.

Figure 4.2 Breeds of sheep and their place of origin.

44 Farming Today

But the importance of each breed varies from one part of the country to another. Figure 4.2 shows where some of the more important breeds of sheep originated.

Figure 4.3 Scottish Blackfaces.

Figure 4.4 Two young Border Leicester ewes (**Society of Border Leicester Sheep Breeders**).

Sheep 45

The identification of sheep breeds is made more difficult by the importance of cross breeding to produce lambs which inherit desirable qualities from both parents. The four main groups are: hill and upland breeds, long-woolled breeds, cross breeds and downland breeds.

The *hill and upland* breeds such as the Scottish Blackface (Figure 4.3), Cheviot and Welsh Mountain are small hardy sheep which can survive harsh conditions and form the largest group in Britain's sheep population. Because of the unfavourable state in which they are reared, few of the lambs grow to slaughter weight in their first summer, so they are sold as 'store' lambs after weaning to be fattened in lowland areas where food is more plentiful. As the ewes get older and less able to survive in the hills, they are 'cast' or sold to lowland farms where they will continue to breed for a few more years. Here they are normally mated to a

Figure 4.5 A Suffolk ram (**Suffolk Sheep Society**).

long-wool breed such as the Border Leicester (Figure 4.4) so called because they are descended from the breeds which were once kept for wool. The aim is to produce *cross-bred* ewes which form the major part of the lowland breeding flock. From their mothers they inherit hardiness and a strong mothering instinct, and from their sire's breed good milking ability, larger size and the ability to produce more lambs. These cross-bred ewes are then mated with a downland ram, such as the Suffolk (Figure 4.5) to produce lambs with a high quality carcass for slaughter.

Figure 4.6 summarises these major groups of sheep and the way in which they are interbred to produce fat lambs for slaughter.

Breeding

All British sheep, except the Dorset Horn, have a clearly defined breeding season in the autumn which is triggered by a decrease in the hours of daylight. From around September the ewes have periods of oestrus or heat which last about 30 hours, and occur every 16 days. Like the cow, the ewe will only mate when she is on heat. The length of pregnancy is 147 days (about five months), so that the lambs are born in the early months of the year and are able to make use of the first grass growth.

The number of lambs born varies from one breed to another, as well as between individuals. Hill breeds have more single lambs, while cross-bred ewes have mainly twins, although triplets and even quads can be born.

The year on the sheep farm

The sheep farmer's year starts around October when the rams are put in with the ewes, although before this he must make sure that all sheep are in good breeding condition. Because only a small percentage of the flock will be on heat at any one time, relatively few rams are needed, the normal

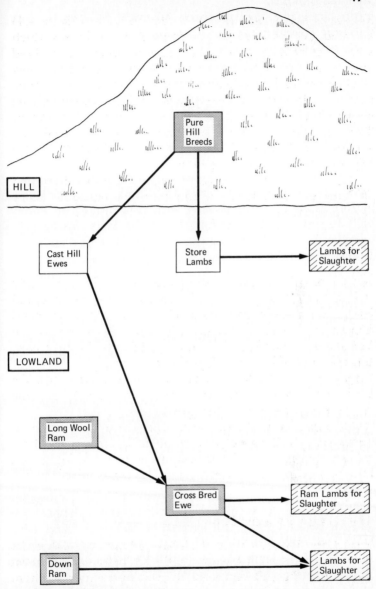

Figure 4.6 Cross breeding of the main groups of British sheep.

ratio being one ram to about 40 ewes. In most lowland farms a coloured marker is strapped to the ram's chest. This will rub onto the ewe's rump during mating so that it is possible to see which ewes have been served by the ram. By changing the marker colour from time to time, the farmer can calculate the approximate lambing date for each ewe, and also identify ewes which 'return', those which keep being mated and are therefore likely to be barren. The rams are usually left with the ewes for three oestrous cycles (48 days) and then taken out.

Through the winter the ewes are kept outside and may need extra hay when the grass runs short, but in the weeks before lambing extra feeding with concentrates or barley will be needed. For lambing itself, the ewes can be kept outside or in, depending on the weather and available housing. A healthy lamb can survive quite severe conditions, but weaker ones will quickly die of exposure, so constant shepherding is needed to identify lambs needing extra attention.

Sometimes the shepherd will try to foster lambs, if for example, a ewe's own lambs have died. This is quite easy immediately after the foster mother has lambed, but becomes progressively more difficult as time goes by. One way of encouraging the ewe to accept the lamb is to skin the dead one and put its coat onto the foster lamb so that it smells right to the ewe. Another is to put the ewe into a stock with her head firmly held, allowing the lamb to suckle without the ewe being able to see or push it away. Usually after a few days the ewe will accept the lamb and they can be turned out together.

In every lambing season there will be some lambs that cannot be reared by their own mothers, and if they are too old for fostering they will have to be reared by hand using powdered milk substitute. The milk is made up with water and fed to the lambs using a bottle or a specially designed milk dispenser.

At some time all the lambs must have their tails docked, and male lambs are castrated. Docking or cutting off the

Sheep 49

lower part of the tail prevents a build up of dung which can act as a breeding ground for flies in the warmer weather. Castration is done because it is thought that older ram lambs will develop a 'taint' or unpleasantly strong taste to their meat. Both docking and castration can be carried out surgically with a sharp knife, or by using a tight rubber band, shortly after the lambs are born. This will cut off the blood supply so that the tissues eventually die and drop off.

The ewe and lambs will be turned out to grass shortly after lambing. The lambs will eat some solid food before they are a month old, but continue to suckle until they are separated from the ewes at weaning time, usually around July or August.

During the summer the ewes are shorn and their fleeces sold. Lambs have very short coats at this time and will not be shorn until the following year. Shearing is a skilled job with the shearer having to hold the sheep with one hand, using the other hand to clip off the fleece in one piece,

Figure 4.7 Shearing.

50 Farming Today

without cutting the ewe. Nowadays, shearing is frequently done by gangs of contract shearers who travel from farm to farm, and electric clippers have largely replaced hand shears (Figure 4.7).

During the summer months, the flock must be dipped or sprayed to prevent 'fly-strike'. This is caused by flies laying eggs in the soiled wool around the tail, which will hatch out into maggots that eat into the flesh of the sheep causing considerable pain and death if left untreated. In addition, in England and Wales there is a period of compulsory dipping in the autumn to prevent sheep scab, which is a skin disease. Dipping is carried out in a specially built bath which must be deep enough to completely immerse the sheep, so that

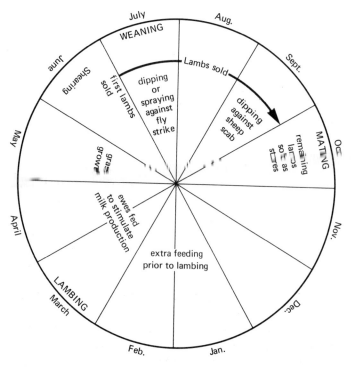

Figure 4.8 The year on a sheep farm.

every part of the fleece is soaked in the dip.

The main activities on a typical sheep farm are summarised in Figure 4.8.

The sale of lambs

Lambs are sold for slaughter as soon as they are fat, but because of differences in rates of growth this may be before weaning or at any time up to a year later. Often lambs which are not fat at the end of the grazing season are sold as 'store' lambs to be fattened by someone else. Technically, after 1st January lambs born the previous spring are called *hoggets*, but they are still sold in the butcher's shop as lamb.

Routine sheep care

Apart from regular inspection of the flock to make sure all the sheep are in good health, there are three operations which have to be carried out as a routine. These are: vaccination, drenching against worms and care of the feet.

Vaccination

Sheep are susceptible to a variety of diseases caused by bacteria, many of which can be prevented by routine vaccination. Ewes are usually given a booster injection shortly before lambing, so that immunity is passed on to the lambs, and they are themselves vaccinated in the first few weeks of life.

Drenching

A number of parasitic round worms which infect the lungs or intestine can be picked up from the grass. These are treated by *drenching*, that is the injection of a measured amount of a liquid drug into the mouth using a specially designed drenching gun (see p.123).

52 Farming Today
Care of the feet
Foot rot is an infection of the hooves causing lameness. Regular trimming of the feet, and in particular cutting away the infected area is essential to prevent this, and the sheep can also be walked through a foot bath containing a chemical to reduce the infection.

CHAPTER FIVE

Pig Farming

Unlike the other animals we have looked at, the pig is an *omnivore*, that is it eats both plant and animal material. Under natural conditions the pig obtains its food by rooting in the ground with its snout and perhaps this is why the pig has a quite unjustified reputation as a dirty animal. When pigs are kept inside they are the cleanest of the farm animals, with separate areas for dunging, feeding and a clean lying area. When kept outside pigs enjoy a mud bath but this is an important way of keeping cool in hot weather because they have no sweat glands. In Canada and the U.S.A. pigs can suffer heat stroke in the hot summers and pig houses are fitted with sprinklers to keep them cool.

In the past pigs were an important part of the cottage economy, often being kept in the back garden and fed on scraps from the house (pig swill) but today the majority of pigs are kept in large herds of 50 or more, under intensive conditions. Pig farms are found throughout the country but because cereals are an important part of the pig's diet, they are most numerous in the arable land in eastern England and also in Northern Ireland.

Breeds of pig

The modern pig is probably descended from a cross between the European wild pig and pigs imported from the Far East. During the eighteenth and nineteenth centuries there were great advances made in the scientific breeding of all types of farm animal and many breeds of pig were developed. Some of these can still be found today, but the only breeds of commercial importance are the Large White and the Landrace.

In recent years a few breeds have been imported in small numbers such as the Piétrain and Duroc from Europe and the Hampshire from America. These breeds have been used to produce hybrid pigs for fattening.

Hybrid pigs

The majority of pigs for fattening are *hybrids* (pigs of mixed breed). This is because a cross between two breeds produces pigs showing *hybrid vigour* (sometimes called *heterosis*) so that they grow faster than either parent. This has led to planned breeding programmes to improve the quality of pigs for fattening. Carefully selected pure-bred pigs are

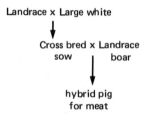

Figure 5.1 Hybrid pig production.

crossed to produce a sow (female pig) with desirable qualities from each parent and she is in turn mated with a pure-bred boar (male pig) to produce hybrid pigs for meat production. One such programme is given in Figure 5.1 and a typical hybrid pig for fattening is shown in Figure 5.2.

In this way, healthy fast-growing piglets are produced which show the best qualities of the pure breeds.

Pig products

Pigs are kept to produce either fresh meat (pork) or cured meat (ham, bacon and gammon) and pig meat is also used

Figure 5.2. A hybrid gilt produced by mating a Large White boar with a Duroc x Landrace sow (**Pig Improvement Company Ltd.**).

to make sausages, tinned meats, pies and other meat products. Pork does not keep as well as other meats and this used to cause a drop in sales during the summer when it was most likely to go 'off'. The introduction of refrigeration and the deep freeze have made the sales of pork more uniform throughout the year.

The methods of curing meat were originally used to preserve meat that could not be used immediately, and today there is no real need to cure pork apart from the fact that people enjoy the different flavour of bacon, gammon and ham.

Bacon and gammon are prepared in a similar way, but come from different parts of the animal (see Figure 5.3). The carcass is first salted after which it can be used at once ('green' bacon) or it may be smoked over smouldering wood for several days producing smoked bacon with a quite different flavour.

Ham is prepared from the hind legs which are removed from the carcass before salting and cured slowly according to local methods, usually involving a sweet pickle of salt

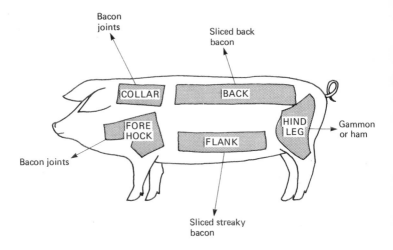

Figure 5.3 Cured joints of pig meat.

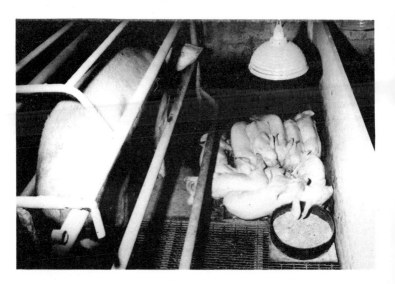

Figure 5.4 A farrowing pen.

and brown sugar. After curing the ham is boiled and usually sold sliced.

Types of pig

Pigs are divided by weight into four main slaughter classes as shown below in Table 5.1.

Grade	Age	Weight	Use
Pork pig	14-18 weeks	120-140 lb (54-63 kg)	Pork
Cutter	18-24 weeks	140-180 lb (63-82 kg)	Larger pork joints
Baconer	22-30 weeks	185-205 lb (84-93 kg)	Bacon and ham
Heavy hogs	26-34 weeks	220-250 lb (100-114 kg)	All purpose, providing some bacon and ham, pork joint and meat for other products

Table 5.1 Slaughter classes of pig.

Most farmers specialise and produce only one grade of pig.

Breeding

Like the other animals we have looked at, the sow will only mate when she is on heat. The gilt (immature female) first comes on heat at about six months of age but is not usually mated until about eight months because litters from younger animals are often small. Sows can be artificially inseminated but this is generally less successful than with cattle, and therefore is not so widely used.

Heat (oestrous) occurs about every 21 days all through the year and at that time between 12 and 20 eggs are released from the ovaries. Most are fertilised but not all will develop and of those that do a small percentage will be still-born so that the number of piglets born alive is usually between ten and twelve.

Farrowing

The length of pregnancy is 115 days (approximately three months three weeks and three days). When the sow is due to farrow she will be wormed and washed with a mange dressing, before being moved to a special farrowing pen. Immediately before farrowing, the udder is washed to prevent the piglets picking up worm eggs or other infections from the teats.

Farrowing normally takes about three to four hours and problems are rare. The piglets are on their feet almost immediately after birth and within a few minutes have found a teat and started feeding. In a few days the litter establishes a 'teat order' with the stronger piglets feeding from the front teats where there is more milk.

The farrowing pen usually has some sort of crate or rail which stops the sow rolling over and crushing the piglets. There should also be a small area (called a 'creep') heated with an infra-red lamp where the piglets lie in the warmth away from their mother, and can be introduced to solid food (Figure 5.4.).

Treatment of the piglets

Piglets have eight needle-sharp eye teeth which are clipped shortly after birth to prevent them from damaging the udder. At the same time male pigs are castrated and the tip of the tail may be docked to prevent tail biting. Piglets reared indoors may lack iron in the diet, causing anaemia and to prevent this they are given iron, either by injection or by mouth.

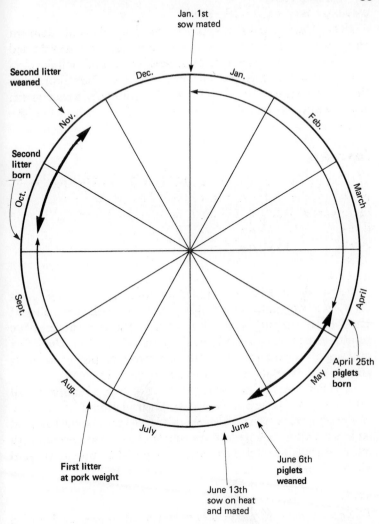

Figure 5.5 A year in the life of a sow.

Weaning

This can take place at any age between about ten days and eight weeks. The advantage of early weaning is that the sow comes back on heat more quickly, and there is less time between litters but the main disadvantage is that the piglets are much more vulnerable to chilling and infections. Early weaning can only be carried out successfully when the piglets are kept in specially designed controlled environment housing. Most farmers wean at around five or six weeks of age, by which time the piglets have developed immunity to many infections and are eating reasonable amounts of solid food.

The sow should come back on heat a week after weaning and will be mated again immediately. In this way it should be possible for each sow to have two litters in slightly less than a year (see Figure 5.5).

Feeding

Feed costs can be as much as 80% of the total cost of pig production which means that the type and cost of food is extremely important. On a large commercial farm the main aim is for the pigs to grow to slaughter weight as quickly as possible, but at the same time not to put on too much fat because most people today prefer lean meat.

Feeding can either be *wet* or *dry*, and is often done automatically. Dry feeding is mainly based on cereals with added protein-rich foods (such as soya or fish meal) plus vitamins and minerals. Some farmers mix their own feed but most buy pelleted food from commercial animal food firms.

Wet foods can include potatoes and other root crops, swill and separated milk (whey), and may be delivered to each pen in a special pipeline which runs through the fattening house. Swill feeding is quite rare now because of its variable make up, as well as the risk of transmitting certain diseases if it is not thoroughly cooked (see also Chapter 10).

Housing

Although the majority of pigs are kept housed, a few farmers specialise in keeping their breeding sows outside, and only bring the young pigs inside for fattening. Pigs kept under these conditions will tend to have less disease problems, but this has to be offset against the cost of the land needed to keep them and the extra labour involved in looking after them. This raises one of the major criticisms of modern farming, so-called 'factory farming'. It is an unfortunate fact of life that larger numbers of animals can be fattened faster and cheaper if they are kept under intensive conditions and this, of course, means cheaper food for the consumer. It is by no means certain that all opponents of factory farming would be happy to pay the higher prices that would follow a return to more 'natural' farming methods.

There are a variety of piggeries available, the design of

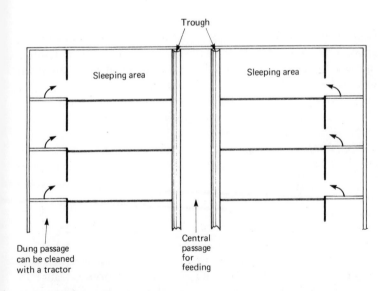

Figure 5.6 Plan of a pig fattening house showing separate sleeping, feeding and dunging areas.

which depends on the type of pig farm, which are:—
- *weaner producers*, who breed piglets and sell them to be fattened by someone else,
- *fatteners,* who buy and fatten weaners
- and *weaner/fatteners* who breed and fatten their own pigs.

As with all farm buildings, the design is important, but particular care is essential for a piggery because of the danger of disease spreading through a large number of animals kept in close quarters. Good ventilation will keep the air moving, but at the same time draughts, which could chill the pigs must be avoided. Cleanliness is essential, so the pens should be simple to clean and they should be designed so that the animals can be easily seen — this means that any pigs showing signs of ill health can be quickly spotted.

Routine on the pig farm

Every day the pigs have to be fed and cleaned out, but the rest of the work is arranged during the week, leaving the weekends free. A pig breeder for example, will arrange matings so that the sow should farrow during the week, but of course, some piglets may still be born over the week-end.

As an example, the weekly routine on one farm where the piglets are fattened up to pork weight is shown below.

MONDAY	Mate sows Treat any litters born over the week-end
TUESDAY	Mate sows Move piglets weaned the previous week into weaner accommodation Disinfect empty pens
WEDNESDAY	Weaning of 6 week-old piglets by taking the sow away from the piglets Move pregnant sows into farrowing pens
THURSDAY	Farrowing

Weekly weighing of weaners and porkers
Porkers grown to slaughter weight collected by lorry
Empty pens disinfected
Feed delivery

FRIDAY   Farrowing
Market day — selling barren sows etc.

Stockmanship

In recent years the fortunes of pig farmers have varied considerably and only the best can continue to make a profit in hard times. The most important differences between successful and unsuccessful pig farmers in probably in *stockmanship*, that is the skill in looking after animals. Because most pigs are kept housed in fairly large groups, disease outbreaks can be even more dangerous than in other animals and an important part of the farmers work is just to watch the stock so that any signs of ill health are seen and treated early.

In addition, between 10 and 20% of baby piglets die in the first few days of life usually by becoming chilled, being crushed by the sow or because of an infection. The greater the care taken of the piglets in these first critical days, the more will survive and this will obviously increase the numbers sold and therefore the profit. It is also important to detect when a sow comes back on heat after weaning, so that she can be mated as soon as possible, and she should be in good condition both at mating and during pregnancy to make sure of a large healthy litter. It is now possible to tell if a sow is pregnant at an early stage by using an ultrasound scanner. There are firms which travel round pig units and use these scanners to check if all the sows are in pig. The earlier the farmer can find out that a sow is not pregnant, the quicker she can be mated again.

In summary, the pig farmer is aiming for large litters with few piglet deaths, and as short an interval between litters as possible. As well as this the pigs should be kept under conditions which will reduce the risks of disease outbreaks, and be fed on a diet to give rapid growth.

CHAPTER SIX

Poultry

The term poultry covers the domestic fowl (chicken), turkeys, ducks and geese, but of these chickens are by far the most important, being kept for egg production (*layers*)

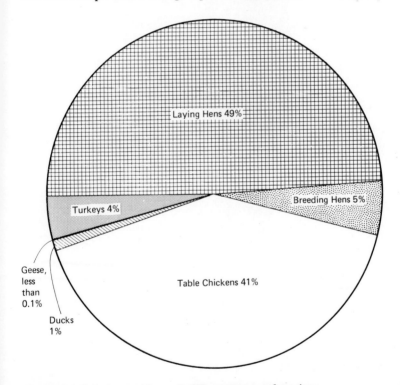

Figure 6.1 Relative numbers of different types of poultry.

66 Farming Today

and for the table (*broilers*). Figure 6.1 shows the relative numbers of different types of poultry kept in the U.K. showing that chickens of one type or another make up about 85% of the national poultry flock.

Figure 6.2 A hybrid laying hen (**ISA Poultry Services Ltd.**).

Breeds of domestic fowl

The pure breeds can be divided into two main groups: *the light breeds* — small birds which lay white eggs and *the heavy breeds* — larger birds laying brown eggs. Nowadays the pure breeds are rarely seen because most poultry farmers keep *hybrids*, that is birds which are produced by a complex programme of selection and cross breeding. This leads to different types of bird suitable for different purposes — laying hens which produce large numbers of high quality eggs and table birds with good body build and a rapid growth rate. A typical modern hybrid laying bird is shown in Figure 6.2.

These hybrids are usually bred by firms specialising in poultry breeding who will sell them to farmers as day-old chicks. In the case of laying birds some farmers buy day-old chicks, rear them to 'point of lay' when the birds are about 18 weeks-old, and then sell them to egg producers.

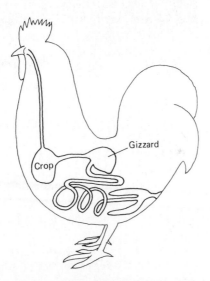

Figure 6.3 The digestive system of the domestic fowl.

Feeding poultry

The hen is an *omnivore*, that is it eats both plant and animal material. Its digestive system is quite different from the animals we have looked at so far, because like all birds, it has no teeth. The food is swallowed whole and stored in the *crop*, before being passed to the *gizzard*, which is a thick muscular sac containing grit. The muscles of the gizzard squeeze the food and the grit together, and this helps to break up the food before it is passed on to the rest of the intestine. Figure 6.3 shows the different parts of the digestive system.

In the past, coarse insoluble grit was fed to poultry but this is less common now because most birds are fed on finely ground meal or pellets which need little or no breaking down. Traditionally farmyard chickens were fed on grain and scraps, but a properly balanced diet must be fed for good growth and egg production. Nearly all farmers buy commercially made feeds, which are designed to meet all the requirements of the bird, but of course this means that there will be different feeds for birds used in egg production and those destined for the table.

Egg laying

Almost all the eggs sold in the shops are unfertilised, that is they were laid despite the fact that the hen had not been mated, and will probably never even have seen a cockerel. The 'blood spots' which are sometimes seen on the yolk are not, as some people think, the start of the development of a chick embryo, but tiny spots of blood which become attached to the yolk of the egg as it is released from the ovary.

A hen has two ovaries but only the left one develops. Figure 6.4 shows the relative positions of the ovary and other parts of the reproductive system, and the part they play in the formation of the egg.

One ovum is released from the ovary each day, and passes

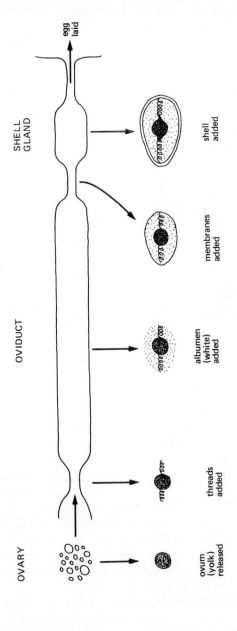

Figure 6.4 The stages in the formation of an egg as it passes along the oviduct.

along the oviduct where it is coated with albumen (egg white), then with the membranes and hard outer shell. The oviduct of a laying hen looks rather like a factory production line with eggs at different stages of formation, with yolks only at the end nearest the ovary, down to a perfectly formed egg ready to be laid. Like all production lines, mistakes can be made and sometimes double yolked eggs are laid, or eggs with soft shells.

There is a rhythmn in egg production with the hen laying eggs on several successive days (called a 'clutch' of eggs) followed by a day when no egg is laid. Some hens have a short clutch (two to three days) while other have much longer ones, from six to as many as one hundred days. Table 6.1 shows some typical laying patterns.

Days	1	2	3	4	5	6	7	8	9	10	11	12	13	14	15	16	17	18	19	20	21
Hen 1	0	0	0		0	0	0		0	0	0		0	0	0		0	0	0		0
Hen 2		0	0		0	0	0		0	0		0	0	0		0	0		0	0	0
Hen 3	0	0	0	0	0		0	0	0	0		0	0	0	0	0		0	0	0	
Hen 4	0	0	0	0	0	0	0	0	0	0	0	0	0	0	0	0	0	0	0	0	0

Table 6.1 Laying pattern of four hens (0 = egg laid).

The length of clutch is an inherited character and it is interesting to note that the modern hen is believed to be descended from the Indian jungle fowl which lays between 25 and 50 eggs in a season. By selecting the best laying birds, poultry breeders have managed to produce the modern hen with a capacity for laying as many as 300 eggs a year.

The egg is produced by the hen to provide food and protection for the chick which will develop if the egg is fertilised. For this reason, the egg is a good source of food, particularly of protein (Figure 6.5).

The shell makes up 11% of the egg and is made mainly of calcium phosphate. To keep the shells hard these minerals used to be fed to hens in the form of oystershell or limestone, but nowadays they are included in the layer's meal.

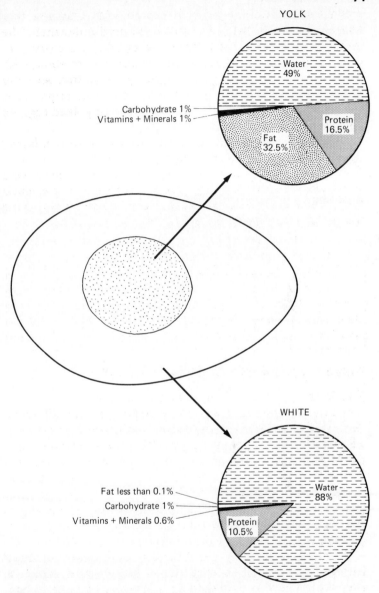

Figure 6.5 Nutritional value of a hen's egg.

72 Farming Today

Shell colour is important because most people in this country prefer brown eggs, although there is no difference in the taste or food value of brown and white eggs, and in some other European countries (such as West Germany) white eggs are more popular. In the past, most breeds laying brown eggs were larger, ate more food and laid fewer eggs compared with the smaller breeds that laid white eggs. In recent years, hybrids have been produced that lay brown eggs but whose performance is as good as those laying white ones.

Breeding

A fertilised egg takes 21 days to hatch, but before it can start to develop it must be kept at a constant warm temperature. Under natural conditions the hen would lay a clutch of eggs over a period of days and then start to brood them by sitting on them. In this way incubation of all the eggs starts together and the chicks hatch out at about the same time. Today almost all eggs are hatched in heated incubators which keep the eggs at constant temperature and humidity.

During incubation the eggs must be turned frequently to stop the developing chick getting stuck to the membrane which lines the shell. In small units this is done twice a day by hand, but in larger ones, it is done automatically. The hen, of course, turns the eggs continuously with her beak and feet during brooding.

The eggs are 'candled' at least once while they are incubating. This involves shining a bright light through the egg so that those developing normally can be separated from infertile ones. The term, candling, dates back to the time when a candle was used to do this.

As the chick develops inside the egg, it uses up the yolk and albumen as food until finally, just before hatching it fills almost the whole shell. When it is ready to hatch, it chips through the shell with a special 'egg tooth' on the beak.

Poultry 73

Broodiness

From time to time laying hens go 'broody' and try to incubate the eggs even though they are unfertilised. While they are doing this, no more eggs are produced, so a tendency to become broody is obviously undesirable in a laying bird. Certain breeds are more likely to go broody than others, and it is important for poultry breeders to try and eliminate this tendency when they are breeding birds for egg production.

Rearing young chicks

If the chicks are being reared for egg production they are sexed after hatching and the males are usually killed. Some breeds have what is called a *sex-linked* down colour, which means that the down of a male bird is a different colour from that of a female. In other cases the chicks are sexed by trained operators who can separate males from females by examining the vent, that is the opening of the intestines and egg canal under the tail.

The chicks are then placed in a brooder which is a cage with a heat source to replace the warmth supplied naturally by the mother hen. As the chicks grow, their soft downy feathers are replaced by adult plumage so that they can maintain their body temperature, and they therefore need less heat. Birds for egg production are sometimes kept outside from about eight weeks-old, but in larger units they are usually inside until they are sold to the egg producer as point-of-lay pullets of about 18-21 weeks.

Egg production

Egg production has become the most intensive, and specialised form of livestock farming, and the number of people involved in it has fallen dramatically. At one stage, on most farms the farmer's wife kept a few hens and sold the eggs. This is relatively uncommon today, when the average egg

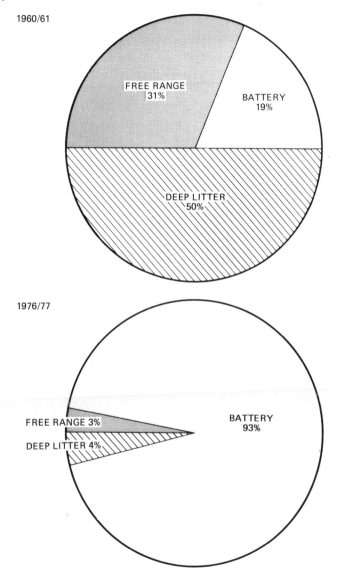

Figure 6.6 Percentage of the laying flock kept on free range, in deep litter or battery cages.

producer is a specialist keeping thousands of hens in battery units.

This change over has been accompanied by a remarkable increase in production. In 1960/61 the average number of eggs laid by a hen in a year was 185, while in 1976/77 it had increased to 243, and has stayed about the same since then.

The three systems for keeping laying hens are: *free range* — where the hens are let out each day, *deep-litter* — where they are kept housed in a poultry shed, on deep litter of straw or wood shavings and *battery cages* — where they are kept confined in battery cages, usually with between three and five birds in each cage.

Figure 6.6 shows how the number of birds kept under each system has changed.

There are several reasons why there has been a move towards housing laying birds. These include:

(1) Saving of land. Very large flocks can be kept in battery cages in a comparatively small space.

(2) Reduction of food wastage.

(3) Control of light. If extra light is not supplied, hens stop laying in the winter as the hours of daylight get shorter. Most commercial flocks have artificial light at all times and the length of the 'day' can be adjusted to encourage the best egg production. This is usually about 17-18 hours.

(4) Temperature control. Free range hens eat more in winter because they need more energy to keep warm. By keeping the hens in well insulated buildings, they keep warm and therefore eat less.

(5) Protection. In some areas foxes can cause considerable losses in free range flocks.

(6) Reduction of labour. A purpose built battery house can allow one person to look after many thousands of hens.

Battery units — a moral problem?
Battery hens have become an emotive issue to many people. Their opponents point to the cruelty of keeping three or more birds in such a confined space where they cannot

76 Farming Today

preen or scratch, but supporters argue that a well-managed system allows the farmer to keep a close eye on the health and well-being of his flock, which is protected from bad weather and foxes. As with many aspects of animal welfare,

Figure 6.7 A battery unit (**Poutry World**).

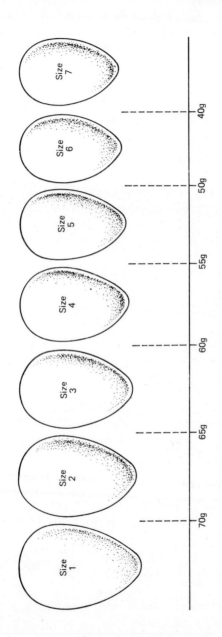

Figure 6.8 Grading eggs by weight.

the question tends, rightly or wrongly, to hinge on simple economics. How much more would people be prepared to pay for non-battery eggs? There are moves to outlaw battery cages and some people predict that there will be a swing away from this method of production as a result of public opinion.

Battery units

These are usually specially built units which contain stacked cages, as shown in Figure 6.7.

The most modern cages have mechanical scrapers for removing the droppings from below the cages, and the birds are fed and watered automatically. As the eggs are laid they roll out of the cage onto a tray in front of the cage, where they can be easily collected. Obviously a system like this is expensive to set up, but once it is working properly one person can care for a large number of birds.

Quality of eggs

Eggs produced from batteries are usually sold to packing stations where they are tested for quality by candling, sorted into sizes according to their weight and packed into cartons for sale in the shops. Figure 6.8 shows the relationship between the size and weight of eggs.

In this way there is control of the quality of the eggs, but many people complain that battery eggs have less taste compared with free range eggs. Certainly there is a marked difference in the yolk colour, because free range hens eat grass which provides a substance called *carotene* giving the free range egg its characteristic deep orange yolk. There is, however, no difference in the food value of free range and battery eggs.

Production of table poultry

The frozen chickens which are found in the deep-freeze

cabinet of every supermarket are called 'broilers' by the poultry trade. These are chickens that are reared in deep litter houses (Figure 6.9) where the feeding and environment are carefully controlled so that they grow rapidly to weights of 1-2 kg (3-5½ lbs) in about seven to ten weeks. The profit on each bird is very small, so it is necessary to have a large number of birds which will grow to a saleable weight as quickly as possible and most broilers are kept in flocks of 50,000 or more.

Figure 6.9 A broiler house, with young chicks (**Poultry World**).

When broiler production first started on a large scale after the second World War, cockerels from egg production strains, were used but now special hybrids, with plump bodies and white flesh are bred.

Turkeys

Turkeys are kept for their meat, and while many are still reared for Christmas, there has been a great increase in the market for turkeys at other times of the year. This is mainly because of the development of smaller turkeys, some weighing as little as 3 kg (7 lbs) as well as to the sale of turkey portions, turkey rolls and other products containing turkey.

Like chickens, newly hatched turkeys need heat, but once they are eight to ten weeks-old they can be moved out of doors for fattening. However, increasing numbers of turkeys are being fattened indoors in conditions similar to those used for broiler chicken production. Most of the birds will be killed between 18 and 26 weeks of age when they weigh between 8-13.5 kg (18-30 lbs).

Most modern turkeys are white feathered although there are still some bronze strains with dark plumage. The most popular breeds have been selected for the breadth of breast, providing extra meat on the carcass but this has made natural mating difficult if not impossible, so that most breeding hens must be fertilised by artificial insemination (AI).

Ducks

Ducks can be kept for the table or for egg production, but the numbers kept for eggs are relatively small because many people dislike their strong taste. There are many different breeds of which the two most important are the Aylesbury for meat and the Khaki Campbell for eggs. Contrary to popular belief, ducks do not have to have access to a pond, and increasing numbers are being fattened indoors for large scale commercial production of table ducks.

Geese

Geese are usually fattened in fairly small numbers for the Christmas market. Although they can be reared indoors they are commonly kept outside, where they feed on grass and, if needed, grain.

CHAPTER SEVEN

Grass

Grass is the main crop in British farming with about two thirds of all agricultural land being in grass. We can divide this into *uncultivated grassland* where natural vegetation exists and *cultivated grassland* that has been sown by man. This is further divided into *permanent pasture* and *leys* (or temporary pasture) which is part of a rotation.

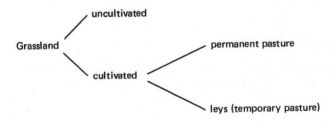

Figure 7.1 Grassland types.

Grass species

A field of grass will contain a large number of different grass species, and other plants. In Britain there are over one hundred grasses, but only a few of these are of much agricultural value. The two most important species are perennial rye grass and Italian rye grass, although others such as Timothy, cocksfoot and meadow fescue are very common.

When a field is sown with grass, the seed mixture used will depend on the use of the grass and the length of the ley.

Farming Today

Figure 7.2 The flowering head of perennial rye grass.

The mixture will be made up of different types of grass and possibly some herbage *legumes,* such as clover. As we shall see later in the chapter, these plants have an important part to play in enriching the soil with nitrogen.

Grassland production

In most parts of the country, grass is not the natural vegetation and would, if left unattended, go through a series of changes of vegetation to become scrub and finally woodland. Because of this, there is a tendency for good productive grassland to deteriorate, so that species of low value to grazing animals start to grow.

The amount of growth and the types of grasses will be affected by a variety of factors, some of which, such as sunlight and altitude, cannot be controlled. However, others, such as drainage, the acidity of the soil, the amount of fertiliser applied, and whether the field is cut or grazed will also affect the grass composition, and by careful management of these factors the field can be maintained in a productive state for many years.

It is important to have a suitable stocking rate for grazing animals, so that the grass growth keeps pace with the amount eaten. This will obviously depend on the species of animal as well as grass type, rainfall and amount of fertiliser applied. On lowland farms, a system of *rotational grazing* is usually followed, where one field is eaten right down and then allowed to rest and regrow while the animals are moved elsewhere. In upland areas there is often a *set stocking rate* when the animals are left to graze over a large area, and are not moved from this. Grass is also grown for hay and silage, and one field may be grazed early in the year, shut off for hay in the early summer and then used for grazing again when the grass has regrown (called the 'aftermath') after haymaking.

Fields may be reseeded either because the grass has deteriorated or because they are part of a rotation, where grass is grown as a 'break' crop, often to rest the land after growing cereals.

There are three main ways of reseeding: *direct reseeding, direct drilling* and *undersowing*. Direct reseeding is done by sowing the seed onto cultivated soil, whereas in direct drilling the existing vegetation is killed with a herbicide such as Paraquat, then the seed is drilled directly into the ground. The most popular method is probably undersowing, where a cereal crop and grass seed are sown together in cultivated land. The cereal grows more rapidly than the grass which tends to reduce the numbers of weeds growing in the grass, and after harvest an established grass crop is left behind.

Application of fertiliser

The most important single nutrient or food stuff for plant growth is *nitrogen* which is needed to make proteins. Plants can only use nitrogen in the form of nitrates, so for maximum growth soil nitrate levels must be kept high. Under natural conditions there is a balance between nitrate taken up by plant roots, and nitrate returned to the soil from the breakdown by bacteria of dead and decaying plants and small animals, together with manure and urine. There are also special '*nitrogen fixing bacteria*' present in the soil, which can convert nitrogen from the air into nitrate in the soil. Clovers and related plants have these bacteria in small lumps or nodules on their roots, and this is why they are often included in grassland mixtures. The natural balance of nitrate in the soil is sometimes called the nitrogen cycle and is summarised in Figure 7.3.

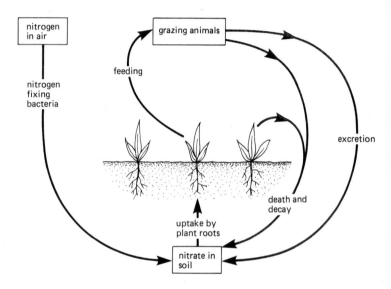

Figure 7.3 A simplified diagram of the nitrogen cycle.

In an intensive grazing system, more nitrogen is being taken out of the soil than can be replaced naturally, so to maintain good grass growth an artificial fertiliser is applied. The reasons for the widespread use of these compared with the more traditional muck spreading are discussed in Chapter 11. The importance of nitrogen as a plant food can easily be seen from Figure 7.4 which shows grass yield after applying different amounts of nitrogen fertiliser.

Plants need other mineral nutrients for healthy growth, but most will be present in large enough amounts in the soil, even after intensive cultivation. However, *potassium*

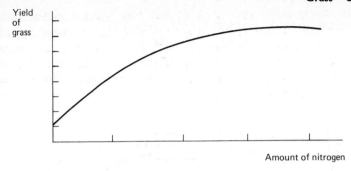

Figure 7.4 Grass yield and nitrogen application.

and *phosphorus* may be required, and some compound fertilisers are a mixture of these two with nitrogen. In addition, *calcium* in the form of lime if often added to acid. soils.

Digestibility

The term *digestibility* is used to describe how much of a feed is digested as it passes through the digestive tract of an animal. The digestibility of grass is usually described in terms of its *D-value* (defined as the percentage of digestible organic material in the dry matter). Grass is most digestible when young and becomes progressively less so as it grows older, mainly because of the greater number of indigestible fibres needed for support as the leaf and stem grow longer. There is a marked increase in fibre once the ears (grass flowers) emerge, and if the grass is allowed to continue growing without grazing or cutting, the digestibility will fall rapidly, so that the more grass there is, the less digestible it becomes, as shown in Figure 7.5.

A value of 65D is normally recommended when grass is cut for hay or silage, as this is a point where there is a good yield without too great a loss of digestibility before flowering. In most parts of the U.K., this would mean cutting in early or mid-May (see Figure 7.5).

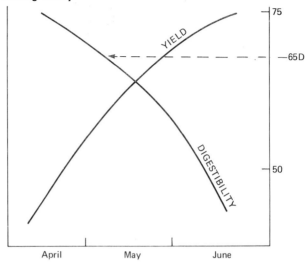

Figure 7.5 Digestibility and yield during the summer.

Digestion in the ruminant

Cattle and sheep are *ruminants*, which means that they have a specially adapted digestive system to deal with the plant material on which they feed. The most important feature of this is the stomach, which is divided into four chambers, called the rumen, reticulum, omasum and abomasum.

Food is briefly chewed and swallowed into the rumen where large numbers of bacteria and other micro-organisms act to break down cellulose and other plant materials which cannot normally be digested. The food is held here for as long as twelve to forty-eight hours, and at intervals some will be regurgitated when the animal is at rest, to be chewed again. This is known as *rumination* or chewing the cud. The food is then swallowed a second time, and passes through the reticulum and omasum before entering the abomasum or true stomach, where digestion continues in a more conventional fashion.

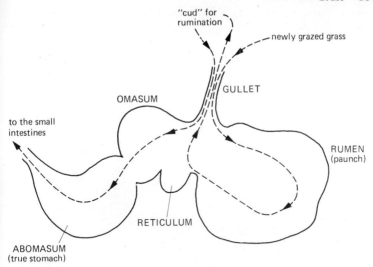

Figure 7.6 The movement of food through the four stomachs of a cow.

Grass conservation

Conservation is the storage of excess grass grown in the summer so that it can be used in the winter. This is done by making hay or silage.

In haymaking, grass is cut with a mower and then dried as quickly as possible by *tedding* and *turning*. Tedding continually mixes up the cut grass so that wet grass is brought up to the surface to dry, while turning means that the whole swath is turned right over so that the side next to the ground is exposed to the air. Once most of the moisture has been lost, the hay is baled and either stacked in the field to dry further or barn dried, using heated air. If bales are stored damp, they will heat up and go mouldy, which reduces the feeding value, and makes it less palatable to the stock, as well as being a possible fire hazard.

To make silage, fresh or partly dried grass is stored in a sealed silo, where air is excluded as far as possible. Under these conditions certain bacteria will ferment the sugars in grass, producing acids which 'pickle' the grass and prevent it from rotting. With either method of conservation, it is important to preserve the grass as quickly as possible once it has been cut, to prevent the loss of nutrients and a decrease in digestibility.

Figure 7.7 A forage harvester picking up grass to be made into silage. (Ford Tractor Operations).

The number of farms making silage is rapidly increasing and on most large farms it is now the major conservation crop. The main advantages of silage making are that it is quicker, and less dependant on prolonged fine spells of weather. In addition, because silage is cut before hay, the crop will be more digestible, and extra cuts can be taken later in the summer. For the small-scale farmer the main disadvantage is that it requires more expensive equipment both for making and for feeding.

CHAPTER EIGHT

Arable Farming

The term 'arable' is usually used to mean any crop except grass, so an arable farm is one with little or no grass. Completely arable farms are fairly uncommon and found mainly in East Anglia, but crop growing is an important part of mixed farming, providing feed for the livestock as well as crops to be sold (sometimes called cash crops).

Although farmers tend to specialise in certain types of farming they will almost certainly grow more than one type of crop because:
1) It is better to rotate the crops so that the same crop is not always grown in the same field.
2) There is less financial risk if several crops are grown — a bad year for cereals may be a good potato year.
3) Growing different crops helps to spread the work load throughout the year. Figure 8.1 shows the times when some common arable crops are sown and harvested.

In the slack summer months the crops will be sprayed and the root crops weeded, but on a mixed farm this will be a busy time with silage and hay making. Winter months will be spent in field cultivations.

Types of crop

The arable crops can be divided as follows:

Cereals
often popularly called corn, of which wheat and barley are the most important.

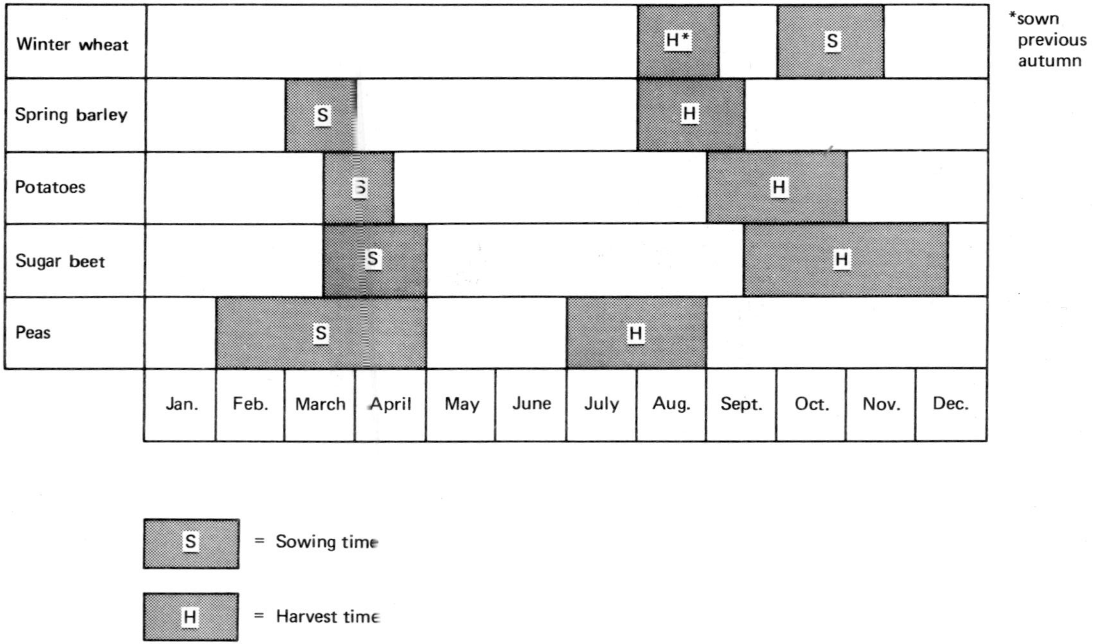

	Jan.	Feb.	March	April	May	June	July	Aug.	Sept.	Oct.	Nov.	Dec.
Winter wheat								H*		S		
Spring barley			S					H				
Potatoes				S						H		
Sugar beet				S							H	
Peas		S					H					

*sown previous autumn

S = Sowing time

H = Harvest time

Figure 8.1 Work involved with some common crops, throughout the year.

Arable Farming

Legumes
the peas and beans which were once important animal feeds, but are less so today.

Root crops
including potatoes (for human consumption) and other roots for animal feeding.

Forage crops
leafy plants grown as a livestock feed.

Oil seed crops
grown for the seed which is crushed to extract the oil.

Figure 8.2 shows the percentage areas of land used to grow different crops in the U.K. More than half the land is in grass and of the land devoted to arable farming, more than three-quarters is used to grow cereals.

Monoculture

Monoculture is when the same crop is grown in the same place year after year. Many people associate this with the formation of 'dust bowls' which happened in America during the 1930's when years of cereal monoculture resulted in the fertile top soil blowing away in dust storms. This is only likely to happen when cereals are repeatedly grown on poor soil, so that the soil becomes light and easily blown away by the wind. The main practical problems with monoculture are a build-up of disease and weeds in the crop, needing increasing amounts of chemicals to control. Because of this most farmers practise some form of rotation.

Crop rotation

As the name implies, the crops are rotated or moved so that the same crop is not always grown in the same place. Throughout the Middle Ages and even later a three-course rotation

94 Farming Today

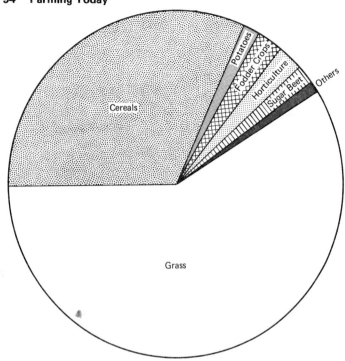

Figure 8.2 Land use in the U.K. (1978).

was followed, where cereals were grown for two years and then in the third year the field was left fallow with no crop. Following the Agricultural Revolution this was largely replaced by other improved rotations, of which the best known is the Norfolk four-course rotation.

This had four main advantages over the three course rotation.

1) It made better use of the land, because none was left fallow.

2) Growing different plants with different needs helped to keep good soil structure and fertility.

3) Rotation of crops helped to stop the build up of pests and disease which normally only affect one sort of crop.

Arable Farming 95

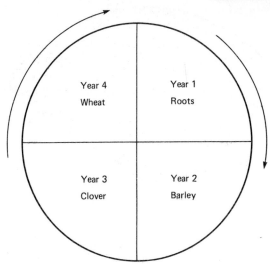

Figure 8.3 The Norfolk four-course rotation.

Figure 8.4 One possible rotation of crops.

4) Growing root crops, which were traditionally hoed by hand, helped to clean the soil of weeds so that the following year cereals were grown in a relatively weed-free soil.

Crop rotations are not followed as strictly as they once were, mainly because of the use of chemicals to kill pests (pesticides) and weeds (herbicides), and the use of artificial fertilisers to maintain soil fertility. In many areas cereals are grown for several years in succession before sowing a 'break' crop, and then returning to cereals. On other farms a rotation is followed, but it will include much more cereal than a traditional pattern. One such rotation is shown in Figure 8.4.

Cereal crops

The cereal crops grown in the U.K. are barley, wheat and oats together with small areas of rye, and maize which can be grown as a forage crop. Figure 8.5 shows the areas of cereal grown, clearly showing the popularity of barley.

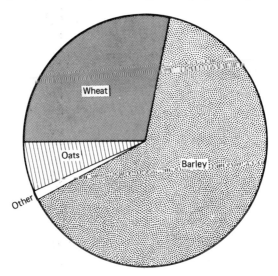

Figure 8.5 Relative areas of cereal grown in the U.K. (1975).

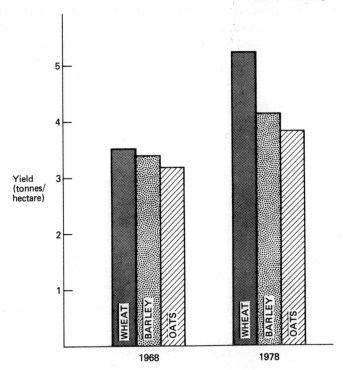

Figure 8.6 Average yields of cereals in 1968 and 1978.

In recent years there have been great increases in yield of cereals (Figure 8.6), mainly due to the development of new high yielding strains of plant.

The main cereals can be easily identified by their ears (Figure 8.7) but it is also possible to separate them at a much earlier stage of growth (Figure 8.8) by looking at the direction in which their leaves twist and the appearance of the auricles, which are small spurs at the point where the leaf meets the stem.

Wheat
Wheat needs good fertile land and is grown mainly in the

98 Farming Today

warmer counties of southern and eastern England where there is a low rainfall. There are a whole range of wheat varieties which can be divided into two main groups, the winter wheats sown in the autumn, and spring wheats sown in the spring. After harvesting, the wheat is either used as an animal feed or milled into flour for bread making and other kinds of baking and cooking.

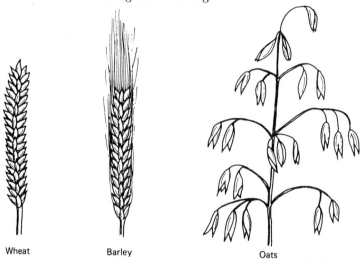

Wheat   Barley   Oats

Figure 8.7 Ears of wheat, barley and oats.

Barley

Barley does not need such good conditions as wheat and can be seen in most parts of the country. It is regarded as being the best cereal for feeding pigs and cattle and most of the crop is used for this, but on suitable land a few farmers specialise in growing malting barley for making whisky and beer. As with wheat, there are winter varieties for autumn sowing and spring varieties which are sown in the spring.

Oats

Oats can tolerate more acid soils as well as colder, wetter conditions so they tend to be found in the northern and

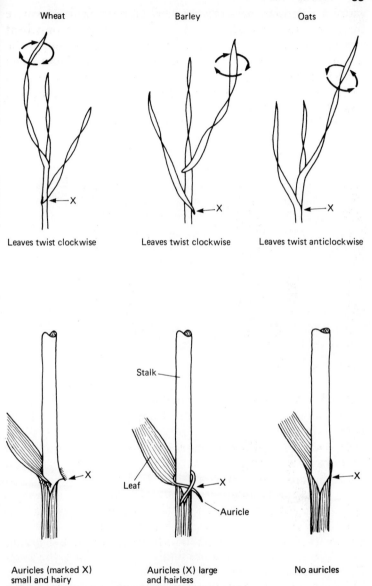

Figure 8.8 Ways of identifying young cereal plants.

western uplands where other cereals will not grow. They are used primarily as an animal feed with a small amount sold as oatmeal for human consumption.

Growing cereals

Before the seed is sown the ground is usually ploughed and cultivated. Cultivation is a term used to cover a range of operations carried out to make the soil in good condition for the seed, and can include: *discing* — to break up the furrows and loosen the soil; *harrowing* — to break down the soil into fine particles and *rolling* — to firm the land and push stones down out of the way of the combine at harvest. More details of the implements used are given in Chapter 9.

The seed is then sowed by drilling into the soil, and after it has sprouted the crop will be sprayed for weed control and against fungus infections. This is usually done with a tractor but in some areas it is now done by low-flying helicopters. As the crop ripens the colour changes from green through to a golden yellow at harvest time, and the ears of wheat and barley droop downwards. Nowadays harvesting is almost always done with a combine harvester (see also Figure 8.9 and Chapter 9).

Cereal diseases

Cereals are attacked by a variety of fungus infections which feed on the plants stunting their growth, and even killing them. Some spread to the plant from the soil and are therefore best controlled by crop rotation, while others may be carried on the seed or by the wind and are treated by seed dressing or spraying. In recent years great progress has been made in breeding strains of cereal which are resistant to some fungus infections.

Weeds

Weeds are a problem in all arable crops because they compete for light, water and soil minerals thus reducing the crop yields. To some extent weeds can be kept down by crop rotation but in cereals the introduction of selective weed-killers has had a dramatic effect, so that fields of corn

Arable Farming 101

dotted with poppies and other weeds are now only a memory. As the name implies, selective weedkillers can be applied to a growing crop where they kill the broad leaved weeds leaving the cereal crop unharmed.

Figure 8.9 A combine harvesting (Massey Ferguson).

Straw

The stalk of a cereal plant is called straw and is used mainly for animal bedding, although both oat and barley straw can be fed to cattle. In the mainly arable areas of the country there are few livestock and therefore only a small market for straw. This makes it hardly worthwhile for the farmer to bale and transport the straw to an area where there is a market, so it may be burnt in the fields after combining. This is often condemned as wasteful, as well as being dangerous if the fires get out of hand, but in an area with a limited market for straw it is hard to see an alternative.

In recent years many larger farms have changed to

102 Farming Today

Figure 8.10 Big round bales (Massey Ferguson).

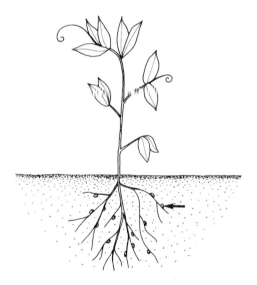

Figure 8.11 A pea plant showing the root nodules.

making big round bales which are either 4 feet (1.2m) or 5 feet (1.5m) in diameter. They are quicker to make and the outer layer makes a weatherproof coat so that they can be left outside until needed.

Legumes

Peas and beans are grown for both human and animal consumption and they make a useful break crop between cereals. All the plants of this type have root nodules containing nitrogen fixing bacteria which enrich the soil with nitrogen, so there is a valuable effect on soil fertility.

Cultivation and drilling of the seed is carried out in a similar way to cereals and the crop is usually harvested either with a combine or cut and shelled by a special machine. With the increasing importance of food processing, special strains of peas and beans have been developed for freezing and canning, and these are often grown under contract to the food factories which are found chiefly in the south and east of England.

Root crops

Root crops are grown for a variety of reasons. Some, such as potatoes and carrots are grown for human consumption, whereas others will be found on mixed farms as animal feedstuffs and sugar beet is grown for sugar manufacture.

In a traditional rotation, frequent hoeing of root crops was an important method of weed control, but the use of weedkillers has made this less important today. They have to some extent been replaced in popularity by the forage crops which are cheaper and easier to grow, although modern machinery for planting and harvesting together with the use of weedkillers has helped to reduce some of the work involved in growing root crops.

Some of the common root crops, together with their uses and the area where they are grown are shown in Figure 8.12.

	Area grown	Use
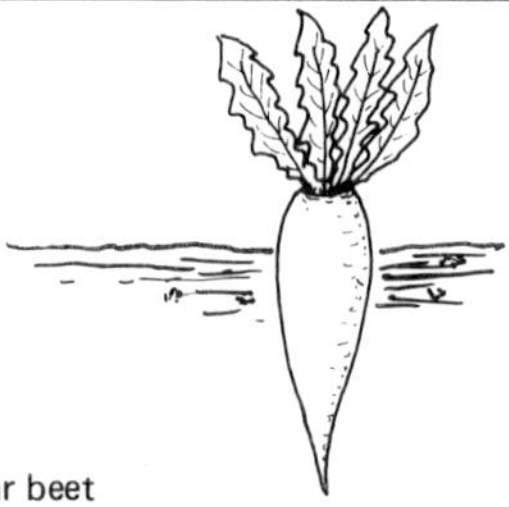 Sugar beet	Mainly in the East (near sugar beet factories)	Sugar manufacture. Tops and pulp (from sugar production) used as animal feed.
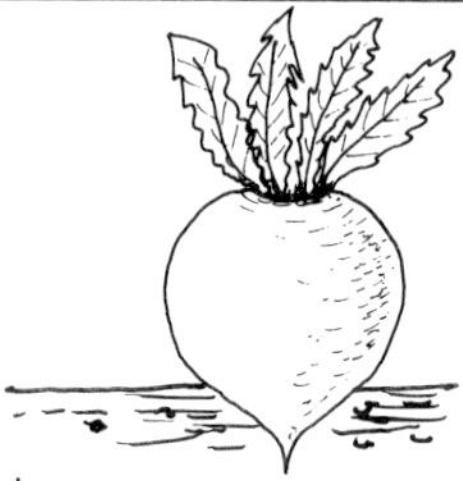 Mangold	South and East (warm dry conditions)	Dairy cows, cattle and sheep feed.
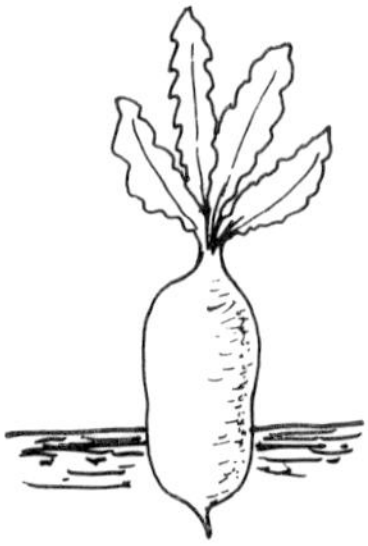 Turnip	North and West (cool damp conditions)	Dairy cows, cattle and sheep. Human consumption.
Swede		

Figure 8.12 Common root crops.

Potatoes which are one of the most important root crops are grown for several different markets:

Early potatoes
These are lifted early before they are mature to give small potatoes with a very thin skin.

Maincrop potatoes
They are left in the ground to reach full maturity before being lifted and stored for sale throughout the winter.

Potatoes for processing
A range of different varieties of potato are grown for manufacture into crisps, chips, dried potatoes and for canning.

Seed potatoes
The production of seed potatoes is a very specialised business because it is important that they should be free from disease. The main disease problems are caused by viruses which are spread by aphids (greenfly) so that most seed potatoes are grown in the cooler areas of the country particularly in Scotland and Northern Ireland, where there are fewer greenfly.

Potatoes are grown from so-called seed potatoes which are small potatoes which are planted in the ground and grow into new plants. The seed potatoes may be chitted (allowed to sprout) before planting (see Figure 8.13).

Figure 8.14 shows how the potato plant grows from the seed potato and new potatoes develop on the roots.

Forage crops

These are crops other than grass whose leaves are used to provide green food for livestock, and in particular for feeding dairy cows through the winter. They can be grazed, cut and fed elsewhere or made into silage. Two of the

106 Farming Today

commonest crops are rape and kale, both members of the cabbage family, but recently maize, which is a cereal, has been grown. The heads which are similar in appearance to corn on the cob will only ripen in the warmer parts of the country, but in other areas the crop is cut green and made into silage for winter feeding.

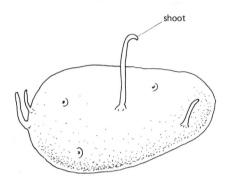

Figure 8.13 A seed potato after chitting (sprouting).

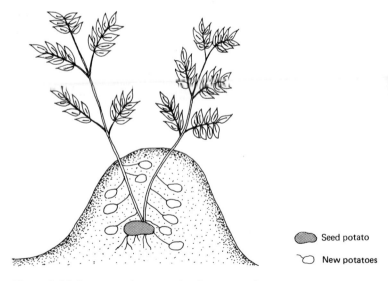

Figure 8.14 A potato plant growing from a seed potato.

Oil seed crops

These are crops which have an oil-rich seed. The only one grown to any great extent in this country is oil seed rape, which has been grown in increasing amounts in recent years as a break crop. After harvesting the seeds are collected and crushed to remove the oil, which is used for cooking oil and margarine, while the 'cake' left behind is used as an animal feed.

Plant pests

The majority of pests which affect crops are insects and they can do damage in three main ways.
1) Biting. The plant can be damaged or even killed by having its leaves or roots eaten.
2) Sucking. Some insects such as the greenfly can suck plant juices.
3) Spread of disease. Many of the sucking insects transmit diseases from one plant to another.

Some pests only affect specific types of crop and can therefore be controlled by crop rotation, but others attack all types of plant and must be treated with chemicals (see also Chapter 11).

CHAPTER NINE

Farm Machinery

One of the most dramatic changes in twentieth century farming has been brought about by the introduction of machinery, and in the first chapter of this book we saw how over the past fifty years, machines have replaced farm labourers, so that in the United Kingdom today farming is highly mechanised with a relatively small labour force. There are now a variety of machines and implements available to the farmer, and in this chapter we shall look at a few of the most important types, starting with the tractor, which has done more than anything else to change the face of farming.

The tractor

The first tractors were produced early this century to replace the horse as a means of pulling farm implements and over the years they have become increasingly powerful and sophisticated. Their usefulness was greatly increased by the development of a *hydraulic system* in the 1930's, for lifting and manoeuvring implements, so that now a tractor can be used to: pull implements — such as balers or trailers; lift implements — using the hydraulic system and drive implements — using the *power-take-off shaft* (p.t.o.) so that implements such as balers can be driven by the tractor while they are being pulled.

The appearance of the tractor has changed markedly since the early days, as can be seen from the illustrations. Figure 9.1 shows the model F tractor which was first manufactured in America in 1917 and 7000 of these tractors

110 Farming Today

were exported to Britain to help boost food production during World War 1. It had metal wheels, no brakes and a gearbox with three forward speeds and one reverse and like all tractors made before the middle of the century was powered by petrol or vaporising oil (paraffin). Figure 9.2 shows a pre-war Fordson and Figure 9.3 a modern general purpose tractor, with a 62 horsepower diesel engine, a gearbox with eight forward speeds and two reverse, a complex hydraulic system and power-take-off shaft, together with a safety cab designed with the driver's comfort as well as safety in mind.

Figure 9.1 A model F tractor, first produced in 1917 **(Ford Tractors).**

Today a wide range are manufactured, ranging from tiny tractors for use in market gardens, orchards or parks to huge tractors, with engines of 300 or more horsepower (Figure 9.4) which would only be found on a few of the largest arable farms.

Farm Machinery 111

Figure 9.2 A pre-war Fordson ploughing **(Ford Tractors)**.

Figure 9.3 A modern general purpose tractor **(Ford Tractors)**.

Figure 9.4 A large tractor, suitable for a big arable farm (**Ford Tractors**).

Figure 9.5 An eight furrow plough **Ransomes, Sims and Jefferies Ltd**).

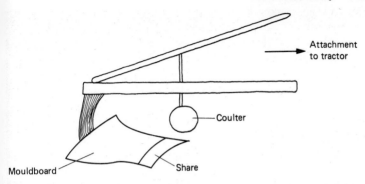

Figure 9.6 The main parts of the plough.

The plough

For centuries farmers have known that to grow good crops the soil must be cultivated by ploughing. The earliest ploughs were shaped pieces of wood drawn by men or oxen which turned over a single furrow, while modern ploughs are tractor drawn, and lifted by the tractor's hydraulic system. Two, three or four furrow ploughs are most common, but the biggest farms with large fields may have up to eight furrow ploughs (Figure 9.5).

Figure 9.6 shows a simplified diagram of the main parts of the plough, but in practice a range of different ploughs are made for different soil types and cultivations.

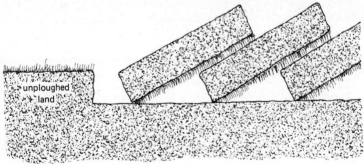

Figure 9.7 The appearance of the land after ploughing.

114 Farming Today

The coulter makes a downward cut into the soil to separate the furrow from the unploughed land. The share makes a sideways cut under the soil and the mouldboard turns the furrow slice over.

Most ploughs are right-handed, that is they turn the furrow over from left to right, but there are also *reversible ploughs* (Figure 9.8). These are really double ploughs, with left handed and right handed parts that can be used in turn. The difference can be seen by comparing the appearance of the furrows in Figure 9.8, where the left handed part of the plough is in use, with the right handed plough shown in Figure 9.5.

Figure 9.8 A four furrow reversible plough **(Ransomes, Sims and Jefferies Ltd).**

Machinery for cultivation

Cultivation is carried out after ploughing to break down the soil into fine particles, between which air and water can pass freely. This gives a good seed bed, providing the seed

Farm Machinery

with suitable conditions for sprouting and early growth.

The effect of cultivation can be seen by looking at the soil in Figure 9.8 where it has been ploughed, and then in Figure 9.9 after it has been cultivated with a disc harrow to break up the lumps of soil. There are many different cultivation methods, including:

rotary cultivators — with blades turned by the p.t.o. shaft to cut through the soil;

disc harrows — sets of discs which turn as they are drawn along (Figure 9.9);

cultivators — with rigid frames fitted with curved tines (spikes) for cutting through the soil;

harrows — with small straight tines for final preparation of the seed bed and covering the seeds;

rollers — which are similar to, though larger than, garden rollers, and are pulled behind tractors to break up lumps of earth, to firm the soil and to bury stones.

Figure 9.9 A disc harrow **(Ransomes, Sims and Jefferies Ltd).**

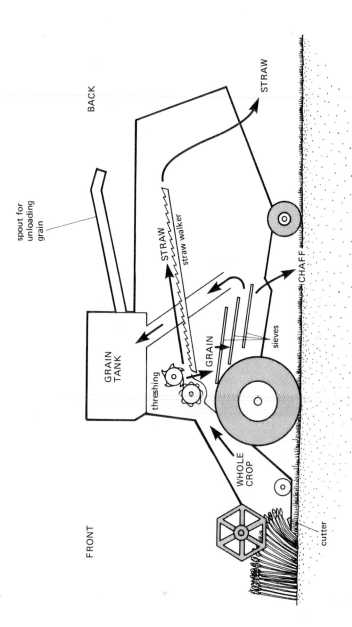

Figure 9.10 A simplified diagram of the working of a combine harvester.

Farm Machinery

The combine harvester

Before the development of the combine harvester there were two stages of harvesting. The ripe corn was cut using a *reaper,* put into stooks and then threshed using a *threshing* machine to separate the grain from the straw. The combine harvester combines both these operations in one machine, hence its name. Figure 9.10 outlines the way the combine works, and this process can be divided into three main stages:

1) cutting. The cutter at the front of the machine cuts the crop just above ground level and it is then passed up to the threshing drum.

2) threshing. The crop passes between rotating drums so that the grain is removed from the straw.

3) separating. The straw moves along a platform, called the straw walker and is dropped out of the back of the machine. The grain and chaff fall down into the sieves which separate them, and the grain is then passed up an elevator into the grain tank for storage, while the chaff falls out onto the ground.

Figure 9.11 A combine harvester **(Massy Ferguson (UK) Ltd).**

Figure 9.12 Two views of a baler **(Massey Ferguson (UK) Ltd)**.

Farm Machinery 119

Figure 9.13 A big round baler (Massey Ferguson (UK) Ltd).

Balers

These are machines which pick up hay or straw that is lying in the field, compress it into bales and then tie them with string. There are two main types of baler, depending on the shape of the bale to be produced: *small square balers*, which produce the traditional rectangular bales and *big round balers*, which produce circular bales either 1.2 metres (4 feet) or 1.5 metres (5 feet) in diameter.

Figure 9.12 shows two views of a square baler. The front view shows hay being picked up and fed into the machine, while the back view shows bales of straw being dropped from the machine and collected with a *bale sledge*. This sledge drags the newly formed bales along after the tractor and baler, until quite a number have collected which are then dropped together in the field so that they can be more easily stacked.

The more recently developed big round baler can be seen in Figure 9.13. The popularity of the two types of baler varies in different parts of the country, but on large farms with suitable equipment for picking up and carrying them, the larger bales are easier to handle compared with stacking and moving large numbers of small bales. As well as this, the outer layer of the round bales makes a weather-proof covering, so that the bales can, if necessary, be stored outside, or even left at the side of the field until they are needed.

CHAPTER TEN

Animal Disease

Farm animals suffer from a wide variety of diseases and these will tend to become more of a problem when large numbers of animals are housed together. To keep stock healthy and free from problems of disease, the farmer needs to consider four main points.

1) *Condition of the animals.* Well fed animals in good condition are much less liable to infection compared with those in poor condition.

2) *Stress.* Any stress, such as a sudden change in housing or feeding will lower resistance to disease.

3) *Buildings.* These should be warm, dry, well ventilated and free from draughts.

4) *Cleanliness.* The buildings should be designed for easy cleaning so that disease is not passed from one batch of animals to another.

It is worth noting at this point that animals which are contented and healthy will grow better than those that are stressed and unsettled, so it is in the best interest of the farmer to keep his animals under good conditions. These may be 'unnatural', as for example, when beef cattle are kept housed, but as far as one can judge they appear quite contented and certainly suffer less stress than when kept outside in harsh weather.

Outbreaks of disease are expensive for the farmer not only because of the cost of treating the sick animal but also because a sick animal is less productive. The example in Figure 10.1 will help to show this — a ewe expecting twin lambs has pneumonia.

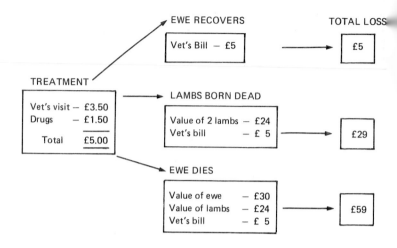

Figure 10.1 The financial loss resulting from a sick ewe.

The possible loss is, in fact, even greater because although the lambs are valued at £12 at birth, they would probably be worth about £35 each when they were ready for slaughter.

Obviously, as far as possible, prevention of disease is better than cure, not only for the well being of the animals but also on strictly economic grounds. One way of reducing the risk of infections is to make sure that all new born animals receive some *colostrum* or first milk, even if they are going to be artificially reared. As explained in earlier chapters, colostrum contains antibodies from the mother which helps to protect the new-born animal from infection in the early weeks of life. As the animal gets older its resistance to some infectious diseases can be increased by vaccination, but for many diseases there is no effective vaccine.

Methods of dosing animals

A sick animal must be dosed with the appropriate medicine as soon as possible. This is normally done by injecting into the muscle or under the skin with a syringe (Figure 10.2), or by *drenching*. This involves the use of a specially designed 'gun' (Figure 10.3) which is used to squirt liquid down the animal's throat.

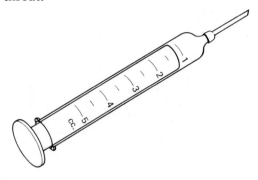

Figure 10.2 A syringe for injection.

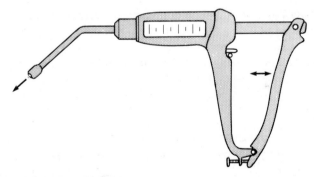

Figure 10.3 A drenching gun.

Small animals such as sheep, can be restrained and dosed by one person, but this is impossible with cattle, where a cattle crush or handling crate has to be used (Figure 10.4).

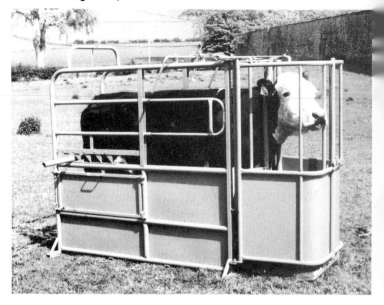

Figure 10.4 A cattle handling crate (**Ritchie Farm Equipment**).

Types of disease
It is obviously impossible to cover the whole range of health problems found on the farm, in one short chapter, so a few of the more important diseases have been chosen and will be looked at under three main headings.
1) Infective diseases
2) Diseases caused by parasites
3) Metabolic disorder.

Infective diseases

These diseases can be passed from one animal to another, and are mostly commonly caused by bacteria (Figure 10.5), or viruses.

Brucellosis
This disease, which causes *abortion* (the premature birth of

a dead calf) in cattle, is extremely infectious, and it can also be transmitted to man, when it causes a disease known as *undulant fever*. A cow will only abort once, but will then act as a carrier of the disease, infecting other animals in the herd. Since 1972 the government has run a compulsory eradication scheme which should eventually make the whole country free from the disease.

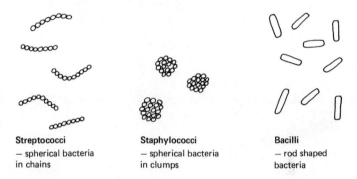

Streptococci
— spherical bacteria in chains

Staphylococci
— spherical bacteria in clumps

Bacilli
— rod shaped bacteria

Figure 10.5 Some types of bacteria as seen under the microscope.

Foot and mouth disease
This highly infectious disease is caused by a virus which affects all cloven-hooved animals (cattle, pigs, sheep and goats) producing lameness and blisters on the mouth. The disease is not normally present in the U.K. and the last major outbreak was in 1967-8. However in March 1981, following cases in Northern France and Jersey, foot and mouth disease was confirmed at a farm on the Isle of Wight, but fortunately did not spread to the British mainland.

In this country outbreaks are controlled by strict isolation of the infected farm together with the compulsory slaughter of all cloven-hooved animals even if they show no signs of ill health. The Ministry of Agriculture will then compensate the farmer for the value of the livestock that have been killed. Since the outbreak of 1967-8 there has been a great deal of discussion about replacing the slaughter policy by one of vaccination against the disease. The rapid control of the Isle of Wight outbreak, together with recent cases of the disease in Europe where vaccination is practised make this change in policy seem less likely now.

Newcastle disease or fowl pest

This is a virus infection in chickens and turkeys which can range from a mild attack with few signs of ill health to very severe with gasping, diarrhoea, paralysis and a high death rate. At one time the disease was subject to a compulsory slaughter policy but this was abandoned when it seemed that this was not controlling the spread of infection while the amount paid in compensation was increasing, and for many years Newcastle disease was controlled by vaccination. In 1981 the vaccination policy was dropped. Any infected birds will be slaughtered but there have been no outbreaks reported for some years.

Swine vesicular disease

This is caused by a virus, producing symptoms very similar to foot and mouth but only affecting pigs. It was first seen in the U.K. in 1972 and has since become a considerable problem. Like foot and mouth disease infected herds must be slaughtered although in this case only the pigs are affected. Most outbreaks have been traced to the feeding of uncooked swill which contained meat from an infected animal, and to prevent this all swill must be boiled for at least an hour before feeding.

Tuberculosis (T.B.)

The main danger of tuberculosis in cattle is that it can be transmitted to humans either by contact with the cows or by drinking unpasteurised milk. In the past a number of different schemes were used to fight the disease and since 1966 the U.K. has been officially declared free from T.B. in cattle. Dairy herds are regularly tested and any cows showing signs of infection are slaughtered. In some parts of the country badgers are thought to act as reservoirs of the disease, causing outbreaks in nearby dairy herds. Because of this, farmers have tried to reduce the numbers of badgers by gassing them.

Other infectious disease

There are, of course, many other diseases which may not be as dangerous as those mentioned above, but which are much more common and are therefore of more importance in the day to day running of a farm. One example is *mastitis* which has been mentioned in the chapter on dairying and which can considerably reduce the milk yield (and therefore the farmer's income) in a cow that appears quite healthy. Other common infections of the larger farm animals are pneumonia, and scouring (diarrhoea) which affects mainly young animals.

Parasitic diseases

A *parasite* is a plant or animal which obtains its food from another plant or animal, called the host. All animals can be affected by parasites which are of two main types, *the external parasites* such as the flea, living outside the body, and the *internal parasites* such as the tapeworm living inside the body. Many internal parasites have a very complicated life history, part of which may be spent inside another animal (called the intermediate host) and during which enormous numbers of new individuals are produced to increase the chances of some surviving to infect a new host. Often, as we shall see, a knowledge of the life cycle can be combined with drug treatment to help control the spread of parasites.

A few of the more common parasitic diseases are listed below.

Sheep scab

This is caused by tiny mites which bite the skin causing irritation and damage to the fleece, and compulsory dipping has been introduced in England and Wales to control its spread (see also Chapter 4).

Ringworm

This is a skin disease in cattle which can be transmitted to man. It is caused by a fungus which produces irritation of the skin and patchy loss of hair.

Liver fluke

This disease affects both cattle and sheep. The fluke is a tiny flatworm which has a complex life cycle involving the snail as an intermediate host, and as snails are more common in damp conditions, most cases of fluke infection are found in the wetter areas of the country, particularly after a warm, wet summer. The life cycle is shown in Figure 10.6.

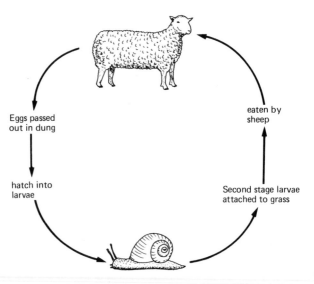

Figure 10.6 The life cycle of the liver fluke.

The adult fluke lives in the bile duct of the liver causing the host to become thin and listless in a mild case, while a severe attack may cause death. Large numbers of eggs pass along the bile duct into the intestine and out with the dung where they hatch out into tiny larvae. These enter the snail where they multiply and change into second stage larvae, which leave the snail to become attached to blades of grass. They are then eaten along with the grass by cattle or sheep, and change to the adult form inhabiting the bile duct.

In damp areas where liver fluke are likely to be a problem the stock should be regularly dosed to kill any adult flukes, while the land should be drained and treated with chemicals to kill snails, so that the life cycle of the parasite is broken.

Roundworms

There are a variety of parasitic roundworms affecting cattle, sheep and pigs which live in different parts of the digestive system. Most have a basically similar life cycle as shown in Figure 10.7.

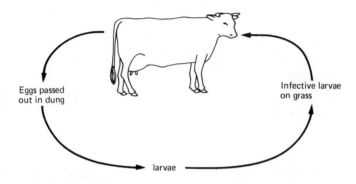

Figure 10.7 The life cycle of a typical parasitic roundworm.

Young animals are much more seriously affected than older animals who develop some immunity, although they still carry worms and can therefore infect the pasture. It has been estimated that one apparently healthy adult sheep can pass out more than eight million eggs in a single day during the spring, which is the time of maximum egg production by the roundworm.

Infections are controlled by regular dosing to kill the worms and by grazing management to prevent a build-up of worms in the pasture. Most worms affect only one type of farm animal so by using a system of grazing rotation where some fields are kept free from sheep for a year and others from cattle, there should always be 'clean' grass available for the young stock. This, of course, helps to explain why most farms are mixed farms. If the fields were never rested from one type of grazing animals there would be a considerable build up of the worm population, and consequent health problems in the stock — hence the old shepherd's saying 'a sheep's worst enemy is another sheep'.

Metabolic disorders

These are not caused by infections but by a disturbance in the *metabolism* or body chemistry of the animal.

Grass staggers

This is caused by a shortage of the mineral, magnesium, in the diet and often occurs when young cattle are turned out onto spring grass, which may be low in magnesium. The animals tremor and stagger and will die unless injected with magnesium. After treatment, they make a complete and immediate recovery.

Milk fever

The mineral, calcium, is present in the body at all times, but is needed in larger amounts than normal by an animal which has just given birth, and started to produce milk. There can be a very sudden fall of calcium levels in the body (because milk has high levels of calcium) which may cause *milk fever*. Although called a fever the temperature is normal but the animal will collapse and die if left untreated. An injection of calcium solution into the vein produces an almost instantaneous recovery.

Pregnancy toxaemia

This is seen in sheep and cattle towards the end of pregnancy. In sheep it is usually called *twin lamb disease* because it is seen in ewes that are expecting more than one lamb and that have not been properly fed. The ewe staggers and becomes blind, but will recover once the lambs are born. The problem is worse when the disease develops some time before lambing so that the lambs may be born dead and the ewe herself may die.

In cattle, it is usually dairy cows with very high milk yields that are affected in the weeks prior to calving but the trouble will clear up as the milk yield falls.

Other metabolic disorders

There are a wide range of other metabolic disorders usually caused by a lack of certain minerals or vitamins in the diet. In the case of minerals there may be an association with a particular soil type that lacks or has low levels of an essential mineral. This means that the crops grown also lack the mineral, as do the animals fed on the crops. Very often the amounts of mineral needed are tiny (the minerals may be called *trace elements*) but severe problems can be caused if they are not present. For example, in some areas the soil is deficient in the mineral cobalt, and the grazing animals suffer from a condition known as 'pine' with loss of appetite and poor growth rate.

CHAPTER ELEVEN

Chemicals

Chemicals are widely used in modern agriculture, and this has been a major factor in increasing food production. Their use, however, has caused concern to some people for a variety of reasons. So in this chapter we shall consider chemicals which may have undesirable effects and not those (such as most drugs) which are generally thought of as completely beneficial. For convenience we can look at the chemicals under three main headings: fertilisers, chemicals used to treat crops and antibiotics and growth promoters.

Fertilisers

Unlike animals, plants can make their own food from simple chemicals during the process of *photosynthesis*. To do this they need: carbon dioxide from the air, water from the soil and sunlight. In addition, they must have *minerals* which are absorbed by the roots from the soil. Under natural conditions these minerals are released as a result of the breakdown of dead plants (see also Chapter 7) but under intensive farming conditions they become depleted and must be replaced with manure or artificial fertiliser. Figure 11.1 summarises the way plants make their food.

There are three main minerals which the plant needs from the soil: nitrogen — to form proteins for growth, phosphorus and potassium — often called potash. Farmers often refer to them by their chemical symbols N (for nitrogen), P (for phosphorus), and K (for potassium). The plant needs other

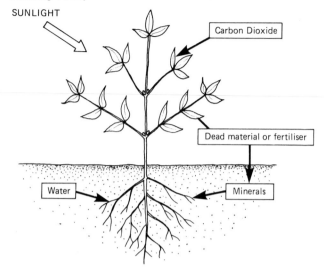

Figure 11.1 Summary of photosynthesis or the way in which green plants make their food.

minerals but because they are required in much smaller amounts they are less likely to be lacking from the soil.

When a crop is harvested the minerals it contains are removed and cannot be returned to the soil in that field, so that some way must be found of replacing them as shown in Figure 11.2. Traditionally this was done by spreading farmyard manure, and although this is still done, the use of 'bag' fertiliser is much more important.

Both manure and chemical fertilisers provide the plant with the same end product — minerals, and the minerals obtained from the breakdown of manure are in no way different from, or better than those out of a bag. The main difference between the two types of fertiliser are in the effects on *organic material* (material in the soil derived from living things). Manure increases the level of organic material, and this has a favourable effect on soil structure. Today much of the manure spread is in the form of 'slurry' which is liquid manure chiefly from pigs and dairy cows where

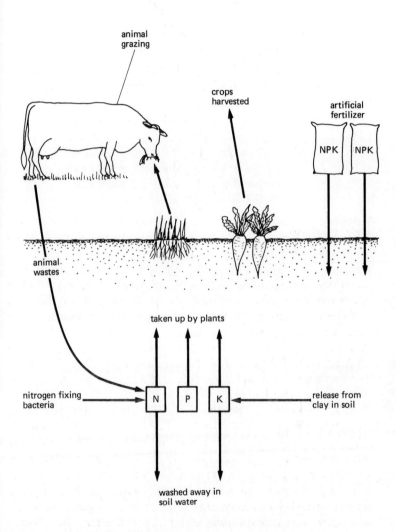

Figure 11.2 Gains and losses of the major soil nutrients.

Table 11.1 Comparison of manure and chemical fertiliser.

	Manure	Chemical fertiliser
Source	Produced on farm	Potash and phosphorus are mined. Nitrogen fertilisers are manufactured by 'fixing' nitrogen from air.
Cost	Free on a mixed farm	Expensive and the cost is rising.
Effect on soil	Increases the amount of organic material which improves the soil structure.	Organic material levels can fall if monoculture is practised on light soils, causing erosion (see Chapter 8).
Chemical composition	Variable	Constant. The farmer can decide how much and in what proportions to add minerals.
Persistence	Slowly broken down releasing chemicals into the soil over a long period.	Dissolve in water, therefore fairly quickly washed out of soil.
Speed of response	Slow	Fast, so can be used to increase growth at times of grass shortage
Pollution problems	Smell may be offensive. Manure polluting waterways causes an increase in numbers of bacteria and a fall in oxygen levels, so that water animals die out. Both problems are worse with slurry.	No smell. Run off fertiliser into streams causes growth of water weeds because of the high mineral levels, which can choke the stream.

straw bedding has not been used. This slurry is usually sprayed onto the fields, but can cause many problems due to its offensive smell, and the possibility of it running off and polluting waterways.

Because bag fertilisers are more flexible most farmers prefer them although manure will also be used on a mixed farm. Each year the use of artificial fertiliser has increased despite rises in cost, but it has been suggested that if costs continue to escalate, levels of fertiliser applied may fall, with a parallel fall in yield.

Artificial fertilisers are usually applied as small granules which can be *broadcast* (spread out) with a fertiliser spreader or drilled into the soil with the seed at sowing time. There are two main types: *straight fertilisers*, containing only one mineral, most commonly nitrogen for use on grassland and *compound fertilisers*, with more than one mineral so that the farmer can select the appropriate levels for the crop being grown. The levels of different minerals are usually given as a ratio called the *plant food ratio*. Some examples are given below in Table 11.2.

Table 11.2 Different types of fertiliser.

Plant food ratio	Composition	Use ratio
2:1:1	2 parts nitrogen 1 part phosphorus 1 part potassium	spring cereals grassland
1:1:1½	1 part nitrogen 1 part phosphorus 1½ parts potassium	potatoes
0:1:1	no nitrogen 1 part phosphorus 1 part potassium	peas

Chemicals used to treat crops

Every year millions of pounds are lost because of damage to crops caused by disease, weed infestation and by pests. As explained in Chapter 8, the problems can be reduced by rotation and cultivation, but major losses can only be prevented by the use of chemicals. We can divide these chemicals into three major groups:

i) chemicals to treat crop disease of which the *fungicides* (which kill fungi) are the most important.

ii) *herbicides* to kill weeds.

iii) *pesticides* which kill pests, of which the *insecticides* are the most important.

Crop disease

Diseases of growing crops can be caused by various organisms, but the two most important are the viruses and the fungi. Viral diseases are difficult to treat and are controlled mainly with insecticides to kill the insects that spread the infection from one plant to another. Fungal infections are controlled either by spraying the growing crop or by seed dressing (dipping the seeds into a fungicide before sowing).

Herbicides

Weeds cause massive reductions in crop yield because they compete for light, water and minerals. There are a bewildering array of herbicides, which can be divided into two main types:

i) *total or non-selective weedkillers*, which destroy all surface vegetation. Paraquat is probably the best known weedkiller of this type.

ii) *selective weedkillers*, which only kill certain types of plant and can be used to kill weeds without damaging the crop.

The times at which they are used will vary between: *pre-sowing* — before the crop is sown, *pre-emergence* — after sowing but before the crop sprouts, or *post-emergence* — on the growing crop. Some are called *residual weedkillers* because they persist in the soil for some time, killing weeds as they emerge.

Chemicals

Pesticides

Although large animals, such as rabbits and woodpigeons can be called pests, the term pesticide is usually used for chemicals which kill invertebrate pests (pests without backbones). Insects are by far the most important group of pests and insecticides are an important group of chemicals in farming. Insects can be killed by contact with the insecticide or by *systemic insecticides,* which are chemicals that are taken up by the plant and kill insects that suck the plant sap.

Problems associated with the use of chemicals on crops

The use of chemicals on crop fields is now almost universal, but despite their undoubted benefits, there are also problems some of which are outlined below.

Persistence: early pesticides, particularly of the organochloride group (including DDT and dieldrin) were broken down very slowly, and tended to be stored in the body. In this way, they could be passed along the food chain, poisoning some of the higher animals. An example of a simple food chain is a barley seed being eaten by a pigeon which is in turn eaten by a fox. If the seed was dressed with dieldrin, this would be eaten by the pigeon, and then in turn by the fox, possibly with fatal results.

Indiscriminate toxicity: chemicals can not only kill the pests, but also other animals which are harmless or even beneficial. Bee colonies have been wiped out when orchards have been sprayed with insecticides, and there are cases where attacks by pests are worse in the years following use of a pesticide because their natural predators, which help to control numbers have also been killed.

Resistance: Repeated use of an insecticide can lead to resistance when the insects are no longer greatly harmed. Insects breed extremely quickly, so that even if only a few can tolerate the effects of an insecticide, they can multiply and produce a resistant strain in a short time.

Aerial drift: Many of these chemicals are applied by spraying, and if it is windy, the spray can drift over field boundaries. This is particularly dangerous with herbicides,

140 Farming Today

which can in this way kill or stunt crops growing in nearby fields.

Antibiotics and growth promoters

These are chemicals which are used in animal husbandry.

Antibiotics are drugs which kill or slow the growth of micro-organisms which cause disease.

Growth promoters are chemicals, many of them antibiotics, which are used to increase the rate of growth of poultry, pigs and cattle.

The first antibiotic to be discovered was penicillin, which revolutionised human and veterinary medicine when it was introduced during the early 1940s. Since then an enormous number of antibiotics have been developed and used in the treatment of disease, but a number of problems have been found of which the most important is resistance. This is similar to the problem already discussed with insecticides, and is summarised in Figure 11.3 below.

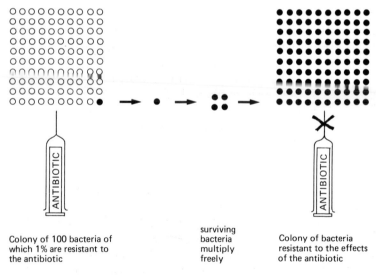

Colony of 100 bacteria of which 1% are resistant to the antibiotic

surviving bacteria multiply freely

Colony of bacteria resistant to the effects of the antibiotic

Figure 11.3 How antibiotic resistance can spread.

How bacteria become resistant is a complex subject, but the most important practical point is that the more often bacteria are exposed to the antibiotic (particularly in small amounts) the more likely they are to become resistant. The problem is complicated because certain bacteria cause disease in both animals and man. Following an alarming increase in drug resistance a committee was set up in the late 1960s under Professor Swann. They recommended that use of antibiotics should be much more strictly controlled, and divided into two groups:
1) antibiotics useful in human and veterinary medicine which can only be obtained on prescription from a doctor or a vet.
2) permitted feed antibiotics, which have no role in disease treatment and can be used as growth promoters.

Despite these controls there is evidence that drug resistance continues to increase particularly in a group of bacteria causing food poisoning and other intestinal infections in man and farm animals.

Growth promoters
These are used to increase the rate of growth in pigs, poultry and cattle. They are either included in the feed, or some types are implanted under the skin of growing cattle.

Many of the growth promoters used for pigs and poultry are either antibiotics or other chemicals with anti-bacterial effects. As previously explained, these are chemicals which have little or no use in medicine, but help to reduce or control some of the infections likely to be found when large numbers of animals are kept together. Another group of promoters include minerals, of which copper is probably the most important. The way in which they work is not clear, but almost all pig foods today contain additional copper.

In cattle, a completely different set of chemicals are used, related to the hormones which occur naturally in the body. These are usually implanted under the skin behind the ear and are slowly released into the bloodstream where they act to produce a faster growth rate with a heavier leaner carcass. However, the use of hormones as growth promoters is likely to be more strictly controlled if EEC proposals are put into action. This follows the Continental boycott of veal, when French consumer groups revealed that certain hormones were being illegally used to treat veal calves. The effect of these hormones is apparently to make the calf retain water, so that the joints are heavier, although of course, the water evaporates during cooking. Claims have also been made that the chemicals used are *carcinogenic* (cancer producing) and cause abnormal development in children.

Although the use of these hormones in veal production seems to be widespread, though illegal, on the Continent, there is no evidence that they have been used in the U.K. where veal production is on a much smaller scale. However, accounts of abuses of the use of growth promoters are bound to make people concerned. They also make us question if the undoubted benefit of using the legally permitted promoters is worth the risks of improper use, and possible contamination with chemicals of meat for human consumption.

CHAPTER TWELVE

Farming and the Common Market

The European Economic Community (EEC) popularly known as the Common Market is based on the Treaty of Rome signed in 1957 by the six original member countries: France; West Germany; Italy; The Netherlands; Belgium and Luxembourg. On January 1st, 1973 the Community was enlarged to nine when The United Kingdom, Ireland and Denmark joined, and then to ten on January 1st, 1981 when Greece became a member. A further expansion will take place within the next few years when Spain and Portugal join the Community.

The Common Agricultural Policy (CAP)

It was decided that there should be a Common Agricultural Policy for all member countries, providing a single market for farm goods with common prices, and free trade between partners. The basic principles of this policy were laid down in the *Treaty of Rome*.

1) *To increase productivity,* that is to make food production more efficient by using modern technology.
2) *To give a fair standard of living to those in farming.*
3) *To stabilise markets* so that there are not large variations in the amounts of food produced or in its price.
4) *To ensure reasonable prices to the consumer,* in other words reasonable prices in the shops.
5) *To ensure security of food supplies* by encouraging home production, thereby reducing imports.

Agriculture is relatively unimportant in the U.K. with

144 Farming Today

Figure 12.1 The EEC.

Farming and the Common Market 145

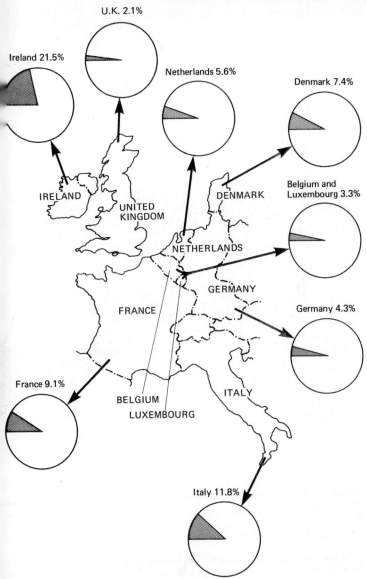

Figure 12.2 Percentage of the population in agriculture (1979).

only about one person in fifty working on the land. By comparison, other Community countries have many more people employed in farming, making agriculture as a source of employment very important to them. Figure 12.2 shows the percentage of the population working in agriculture in the EEC members states in 1979.

However the number of people employed in farming has fallen and is likely to continue to do so, as people move away from the land into the cities.

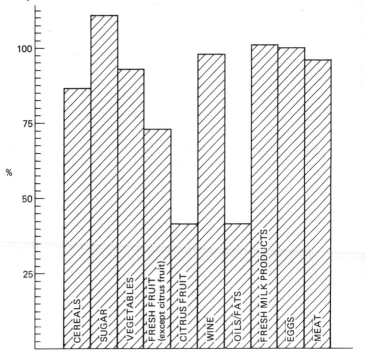

Figure 12.3 The degree of self-sufficiency of the nine member states in various products (1976/1977).

A figure of 100% means that the Community produces, exactly the amount it needs. If the figure is over 100%, the Community has produced more than it needs, creating a surplus.

Farming and the Common Market

Because of the relatively cool climate and high wages, much of Europe is an expensive place to produce food compared with other parts of the world. The Community is however committed to keeping a strong agricultural industry for various reasons which include: *Self-sufficiency:* so that Europe is not completely dependent on imported food, of which the supply may be affected by international crisis or war. *To provide employment in the country* which will help to slow down the migration of people from the country into already overcrowded towns and cities. Figure 12.3 shows the degree of self-sufficiency within the Community of various products.

To stop competition, imports into the Community are taxed so that they do not undercut the more expensive Community products.

Food prices in the U.K.

Before the U.K. joined the Community in 1973, we had enjoyed a 'cheap food' policy for many years. This was due partly to cheap imports from the Commonwealth and partly to the policy of paying subsidies to farmers and keeping food prices low. This contrasted with the Community policy which discourages imports and keeps shop prices high.

The two policies are compared in Figure 12.4, using an imaginary example of the money a farmer received from the sale of a joint of meat. Note that in each case a proportion of the shop price goes to middlemen and not to the farmer.

Policy A is typical of U.K. farming before we joined the Community. The price in the shop is relatively low, but the farmer receives a cash payment (subsidy) to help increase the farm income. *Policy B* is the one adopted by the Community and here the shop price is kept high, so that the farmer gets a reasonable price without a subsidy. The way in which the shop price is kept high is explained in the next section (How the CAP works).

Both these policies give the farmer the same amount of money, but in one case the whole cost is borne by the consumer whereas in the other the government makes a payment.

148 Farming Today

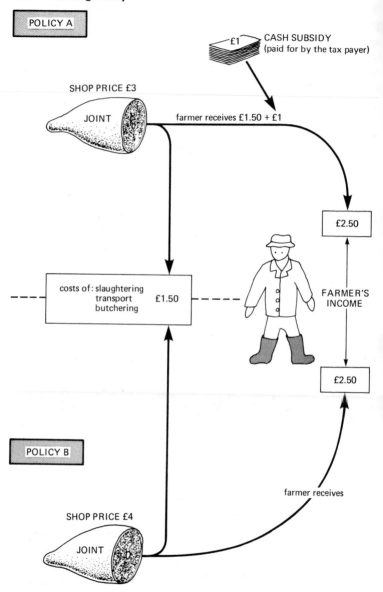

Figure 12.4 Two different farm policies.

Because prices in the Community were high, there was a gradual rise in food costs in the first few years of our membership, and to cushion this, there was a transitional period so that the U.K. did not fully adapt to the Community policy until 1978, five years after joining.

How the CAP works

Every year the EEC's Council of Ministers (made up of farm ministers from each country) meets to set farm prices for the coming year. The way in which they do this varies from one product to another but for about three-quarters of them, two prices are set: the *target price* which is the price that they expect the product to sell for and the *intervention price* which is the lowest price that the product can be sold for, and which is a fixed proportion of the target price. When the price falls to this level, the food is bought by government-run intervention agencies, to stop any further fall in price and is stored, forming the notorious 'lakes' and 'mountains'.

The cost of this policy is met from a Community budget into which each member state pays, and at present the CAP takes up about three-quarters of the budget. In the past few years there has been a lot of disagreement between the U.K. and other Community members because our net contribution to the budget is much greater than theirs. In simple terms, there are two main reasons why this should be so. Compared with other member states we have a relatively small farming sector and therefore get less money out of the CAP. We import more food from outside the Community which is taxed, and the taxes paid into the Community budget, so that we contribute more money.

The green pound

Farm prices are set each year in an artificial currency called *units of account,* and these are changed into the currency of each member state at a fixed rate called the 'green rate'. This means that the French have a 'green franc', the Germans a 'green mark' and we have a 'green pound' for fixing farm prices.

Lakes and mountains

These are the popular names for the large stores of food that have been sold into intervention, although in fact many of them are much smaller than people imagine (Figure 12.5). However the cost of storing these surpluses (particularly those which need refrigeration or freezing) and then disposing of them can be high, and this is an important part of the total cost of the CAP.

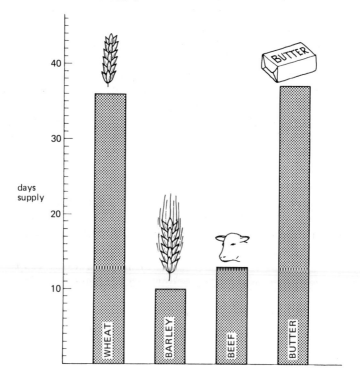

Figure 12.5 The size of some EEC surpluses showing the number of days supply held in intervention (October 1980).

If large surpluses do occur, there are various ways of getting rid of them:

1) *Exporting outside the EEC.* Because world prices tend to

Farming and the Common Market 151

be lower, this will have to be subsidised and may well be unpopular, as when cheap butter was sold to Russia.

2) *Subsiding certain uses* such as providing schools with cheap or free milk.

3) *Aid to less developed countries.* This is done on a small scale, but in the long term it is better to help these countries to produce their own food rather than for them to be given surpluses from richer countries. The other problem is that the surplus foods may not fit into the traditional diet of the country to which they are sent.

To many people, the obvious answer is to sell the products cheaply within the EEC but although the shopper would benefit from this, the farmer would suffer and may even be forced out of business which would have a harmful effect on future food production. The only certain method of stopping mountains building up is to cut down on the production of these products, but in practice this seems hard to do.

The future of the CAP

As we have seen, at present, the CAP takes up three-quarters of the Community budget, leaving relatively little to spend on other important areas such as industrial development. It has been calculated that spending by the EEC is likely to be greater than the whole budget within the next few years and this threat has made it very important that the policy is reformed in some way. The main problem here is to get all the member states to agree as to how the policy should be reformed. Some countries like the U.K. have been pressing for changes for some time, whereas the French, who get a more favourable deal from the present policy are reluctant to support reforms.

The problem will be made worse when the three new countries join the Community. Greece has already become a member (January 1981) and Spain and Portugal will join within the next few years. These are all relatively poor countries who will need some economic aid, and who have large numbers of people employed in farming. The largest

152 Farming Today

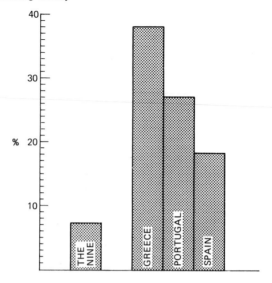

Figure 12.6 The percentage of the population working in agriculture for the nine member states compared with Greece, Portugal and Spain (1979 figures).

country, Spain, will increase the farming population of the Community by one third, and Figure 12.6 shows the much higher percentage of the population working in agriculture in the three new countries compared to the original nine.

Obviously this is going to increase the cost of the CAP at a time when, as we have already seen the Community budget is almost exhausted, so that reform of the farm policy becomes even more important. However it should be remembered that the CAP is a vital part of the Community policy and that its survival in some form is probably essential if the EEC itself is to survive.

Index

Aberdeen Angus 13, 42
Animal welfare 35, 61, 75, 121
Antibiotics 140
Artificial insemination 16, 57, 80
Ayrshire 13

Bacon 54
Bacteria 125
Balers 102, 119
Barley 98
Barley beef 38
Battery hens 75
Beef:
    breeds 32
    carcass 31
    systems of production 37
Breeds:
    beef 32
    dairy 13
    domestic fowl 69
    pig 53
    sheep 41
Broilers 78
Broodiness 73
Brucellosis 124

Calf 18
    rearing 34
Castration 36, 49, 58
Cereals 91, 96-101
    disease 100
    identification 97
    yields 97
Charolais 33
Chick rearing 73
Colostrum 18, 34, 122
Combine harvester 100, 117

Common Agricultural Policy (CAP) 143-152
Crop rotation 93
Cultivations 114

Dairy cows:
    breeds 13
    breeding 16
    feeding 20
    numbers 12
Dairy farming 20, 26
Dairy parlour 23
Digestibility 87
Digestive systems:
    hen 67
    ruminant 88
Dipping 50
Domestic fowl:
    breeds 69
    breeding 72
Dosing 123
Ducks 80

Eggs:
    grading 78
    incubation 72
    laying 68
    nutritional value 71
    production 73
European Economic Community 144

Factory Farming 61
    (see also Animal Welfare)
Farm:
    output 1
    size 5
    types 5
    workers 4

Farrowing 58
Fertilisers 85, 133
Food surpluses 150
Foot and Mouth Disease 125
Forage Crops 93, 105
Friesian 13
Fungicide 138

Gammon 54
Geese 80
Grass:
    conservation 89
    growth 84, 87
    reseeding 85
    species 83
    staggers 130
Grassland 83
Green pound 149
Growth promoters 141

Ham 54
Hay 87, 89
Herbicides 138
Hereford 32
Housing livestock 61, 75, 121
Hybrid:
    pigs 54
    poultry 67

Infective disease 124
Insecticide 139

Jersey 13

Lactation 18
Lambing 48
Land use 94
Legumes 93, 103
Lime 87
Liver fluke 128

Mangold 104
Manure 136
Mastitis 19, 127
Mechanisation 4, 109
Metabolic disorders 130
Milk:
    composition 16
    fever 130
    hygiene 26
    yield 11, 19

Milking 24
Milk Marketing Board 29
Monoculture 93

Newcastle disease 126
Nitrogen 85, 133
    cycle 86

Oats 98
Oestrous (heat):
    cow 16
    sheep 46
    pig 57
Oil seed crops 93, 107

Parasites 127
Pesticides 139
Phosphorous 87, 133
Photosynthesis 134
Pig:
    breeds 53
    breeding 57
    farming 62
    feeding 60
    housing 61
    products 54
    slaughter classes 57
Plant pests 107
Plough 113
Potassium 86, 113
Potatoes 105
Poultry:
    broilers 65
    feeding 68
    layers 65
Pregnancy toxaemia 130

Ringworm 127
Root crops 93, 103
Rotations 93
Roundworms 51, 129
Rumination 88

Shearing 49
Sheep:
    breeds 41
    breeding 46
    farming 46
    numbers 42
    scab 127

Silage 87, 89
Simmental 33
Stockmanship 63
Straw 101
Suckler herds 35, 39
Sugar beet 104
Swede 104
Swine Vesicular disease 126

Tractor 4, 109
Treaty of Rome 143

Tuberculosis 126
Turkeys 80
Turnip 104

Veal 35

Weeds 100
Weedkillers — see Herbicides
Wheat 97
Wool 41